THE STORY OF THE EARTH IN 25 ROCKS

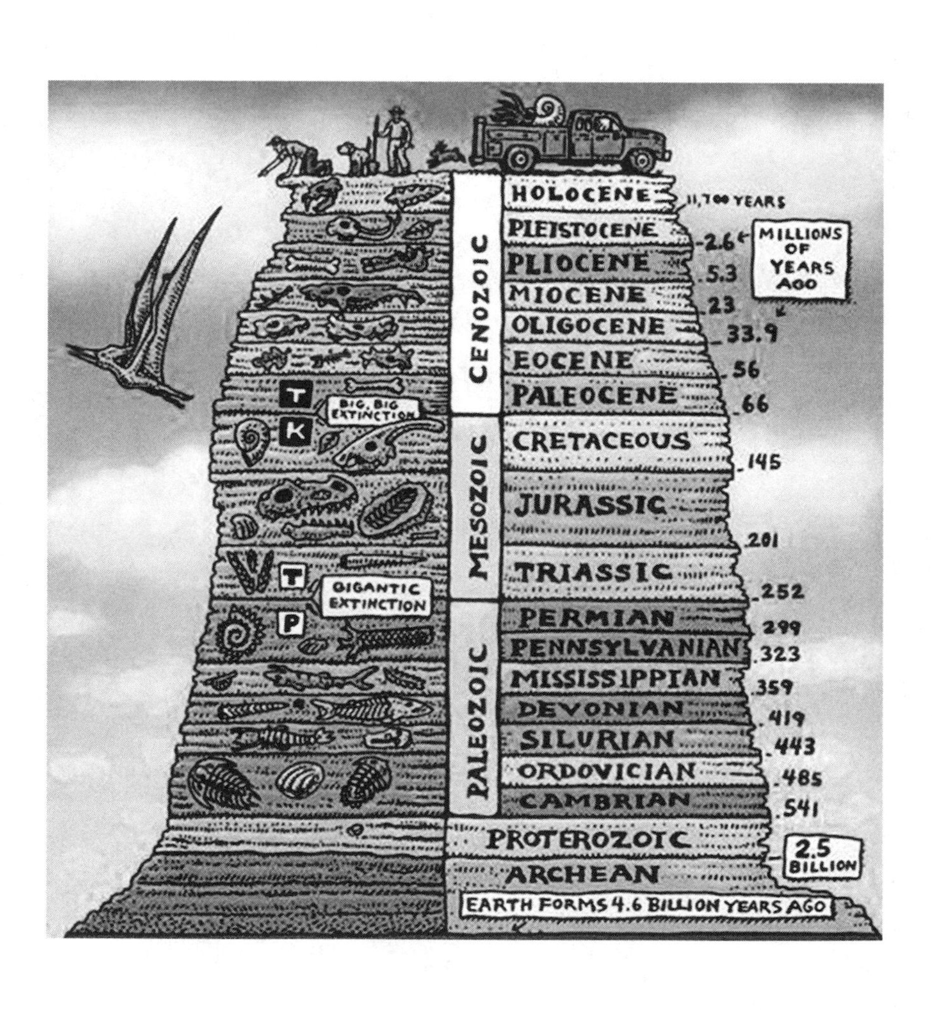
HOLOCENE
11,700 YEARS
PLEISTOCENE
2.6
MILLIONS OF YEARS AGO
PLIOCENE
5.3
MIOCENE
23
OLIGOCENE
33.9
EOCENE
56
PALEOCENE
66
CENOZOIC
BIG, BIG EXTINCTION
T
K
CRETACEOUS
145
JURASSIC
201
MESOZOIC
TRIASSIC
252
T
GIGANTIC EXTINCTION
P
PERMIAN
299
PENNSYLVANIAN
323
MISSISSIPPIAN
359
DEVONIAN
419
SILURIAN
443
ORDOVICIAN
485
CAMBRIAN
541
PALEOZOIC
PROTEROZOIC
2.5 BILLION
ARCHEAN
EARTH FORMS 4.6 BILLION YEARS AGO

THE STORY OF THE EARTH in 25 ROCKS

TALES OF IMPORTANT GEOLOGICAL PUZZLES AND THE PEOPLE WHO SOLVED THEM

DONALD R. PROTHERO

COLUMBIA UNIVERSITY PRESS NEW YORK

COLUMBIA UNIVERSITY PRESS
Publishers Since 1893
New York Chichester, West Sussex

cup.columbia.edu

Library of Congress Cataloging-in-Publication Data
Names: Prothero, Donald R., author.
Title: The story of the Earth in 25 rocks : tales of important geological puzzles and the people who solved them / Donald R. Prothero.
Description: New York : Columbia University Press, [2018] | Includes bibliographical references and index.
Identifiers: LCCN 2017025802 | ISBN 9780231182607 (cloth : alk. paper) | ISBN 9780231544276 (e-book)
Subjects: LCSH: Geology—Popular works. | Geologists. | Earth sciences—Popular works. | Earth (Planet)—Popular works.
Classification: LCC QE31 .P76 2018 | DDC 551—dc23
LC record available at https://lccn.loc.gov/2017025802

Columbia University Press books are printed on permanent and durable acid-free paper.
Printed in the United States of America

Cover design: Julia Kushnirsky
Cover photos by (*top to bottom*): Graeme Churchard; C. T. Lee; McGhiever; J. Valley
Frontispiece: Courtesy of Ray Troll

I DEDICATE THIS BOOK TO THE AMAZING GEOLOGISTS WHO MADE THESE DISCOVERIES, ESPECIALLY THOSE WHO INSPIRED ME IN MY CAREER. THESE INCLUDE:

ALFRED WEGENER
JAMES HUTTON
CHARLES LYELL
WILLIAM SMITH
ARTHUR HOLMES
GERRY WASSERBURG
GENE SHOEMAKER
CLAIR PATTERSON
CHARLES D. WALCOTT
JOE KIRSCHVINK
THOMAS HENRY HUXLEY
ALAN COX
G. BRENT DALRYMPLE
KEN HSÜ
GARY ERNST
BILL RYAN
WALTER MUNK
LOUIS AGASSIZ

CONTENTS

PREFACE

Every rock or fossil tells a story. To most people, a rock is just a rock, but to a skilled geologist, a rock is a clue full of valuable evidence that can be read clearly—if you only know how. I often tell students that geology is like the TV series *CSI*. Geologists and paleontologists act as forensic detectives, piecing together subtle pieces of evidence to reconstruct a "crime scene" from the past—often in incredible detail.

Following the style of *The Story of Life in 25 Fossils*, I have tried to write an accurate but engaging book that is readable and entertaining for lay audiences as well as professionals as a logical sequel. As in *The Story of Life in 25 Fossils*, each chapter focuses on a particular rock or famous outcrop or important geologic phenomenon. I relate some of the fascinating historical and cultural background that underlies the significance of these rocks or phenomena, and how they changed the way people thought about the earth or about how certain of the earth's processes work. In addition, I have tried to weave in the story of the fascinating people who made these discoveries, and how they made them. In most cases, understanding came gradually, with a number of smaller, unexplained discoveries acting as individual pieces of a puzzle. The picture eventually became clear as the pieces were fit together. This theme of unexplained puzzle pieces and their resolution will be found in many of the chapters.

ACKNOWLEDGMENTS

I thank my supportive editor, Patrick Fitzgerald, for suggesting the idea and making many important contributions. I thank Kathryn Jorge at Columbia University Press for overseeing the production of the book and Ben Kolstad at Cenveo for managing the production. I thank Greg Retallack and Nick Fraser for their perceptive comments and suggestions and Paul Hoffman for comments on chapter 16.

I thank my son, Erik Prothero, for editing and redrawing many of the illustrations in Illustrator and Photoshop. I thank many different people for permission to use their images as credited in the captions.

Finally, I thank my loving and supportive family for all their help: my sons Erik, Zachary, and Theodore, and my amazing wife Dr. Teresa LeVelle.

THE STORY OF THE EARTH IN 25 ROCKS

01 | VULCAN'S WRATH: THE ERUPTION OF VESUVIUS

VOLCANIC TUFF

Live in danger. Build your cities on the slopes of Vesuvius.

—FRIEDRICH NIETZSCHE

FIRE OF THE GODS

The eruption of a volcano can be a terrifying event. In ancient times, and even today in many different cultures, volcanoes were seen as a sign of the wrath of the gods or a punishment for transgressing divine decrees. Their enormous power, noise, and potential for destruction made volcanoes more feared than any other geologic event except earthquakes. The Romans thought that the fires of the Mount Etna volcano in Sicily were due to the forges of the god of fire, Vulcan (Hephaistos to the Greeks). He used the heat of the underworld to hammer out armor, metalwork, and weapons for the gods (including the thunderbolts thrown by Jupiter/Zeus). When eruptions occurred, it was said that Vulcan was angry because his wife Venus (Aphrodite to the Greeks) had cheated on him. The Romans considered Mount Vesuvius on the Bay of Naples sacred to Hercules (Herakles to the Greeks), and some scholars think that the name "Vesuvius" is derived from the Greek for "son of Zeus" (as Herakles was Zeus's son).

Yet one of the first truly scientific descriptions and insights into what a volcano is and how it erupts came from ancient times. In some senses, the eruption of Vesuvius in 79 C.E. can be considered the beginning of our modern understanding of Earth and the event that led to the birth of geology as a science.

At the time, the towns around Mount Vesuvius were prosperous and growing. The Bay of Naples supported a large fishing industry, and wine grapes were grown in many places, including the slopes of Vesuvius itself. Then, as now, the volcanic soils around Vesuvius were too rich to be ignored, and they were valued for growing crops and vines of wine grapes. The Roman emperors had a large villa out on the nearby island of Capri, and many other powerful Romans had homes in the region. Pompeii was a large city with a population of over 20,000, and there were many smaller communities in the region.

Vesuvius had not erupted since 217 B.C.E., so most Romans thought it was extinct. Yet there had been many earthquakes in the 17 years since the great earthquake of 62 C.E. that had destroyed much of Pompeii, Herculaneum, and Neapolis (Naples). As early as 30 B.C.E., the Greek historian Diodorus Siculus described the Campanian plain as "fiery" (*Phlegrean*) because Vesuvius showed signs of the fires that had burned long ago.

Figure 1.1 ▲
The 1944 eruption of Mount Vesuvius. (Courtesy of Wikimedia Commons)

By August of 79 C.E., earthquakes had become more frequent, and many of the wells and springs in the area had dried up. August 23 was the Vulcanalia, the feast day of Vulcan, which the Romans celebrated every year. Ironically, the next day Vulcan replied to his worshippers with a huge explosion that darkened the skies with ash and a rain of pumice that lasted 20 hours (figure 1.1). Some of the people of Herculaneum and Pompeii evacuated immediately, but there were many left behind, unwilling to leave or unable to do so because there were not enough boats in the harbor and the roads were clogged with traffic and almost 2.8 meters (9 feet) of fallen ash and pumice. Not only was it hard to escape, but it was even hard to breathe with the ash, which choked the lungs of people and animals. But this was only the warm-up. A day later, Vesuvius spewed out many *nuées ardentes* ("glowing clouds" in French) or pyroclastic flows. These superheated (up to 850°C, or 1,560°F) mixtures of volcanic gases and ash roared down the mountain slope at 160 kilometers per hour (100 miles per hour), incinerating everything in their path. They buried Herculaneum under tens of meters of volcanic deposits known as tuff.

A HISTORIAN OF DISASTER

The eyewitnesses of the eruption mostly died or left no written records, so their thoughts have been lost in the mists of history. Fortunately, we do have one excellent eyewitness account, written by the historian Pliny the Younger. He was 17 at the time, fleeing with his family in a boat to the town of Misenum, across the bay 35 kilometers (22 miles) from the volcano. In a letter to his friend, the famous historian Cornelius Tacitus, the younger man described how his 56-year-old uncle Pliny the Elder, one of Rome's leading admirals, scholars, and naturalists, decided to take a boat closer to the mountain to rescue his friends. It has been one of my favorite accounts of any eruption ever since I first read it in the original in my high school Latin class:

> My dear Tacitus,
>
> You ask me to write you something about the death of my uncle so that the account you transmit to posterity is as reliable as possible. I am grateful to you, for I see that his death will be remembered forever if you treat it [in your

Histories]. He perished in a devastation of the loveliest of lands, in a memorable disaster shared by peoples and cities, but this will be a kind of eternal life for him. Although he wrote a great number of enduring works himself, the imperishable nature of your writings will add a great deal to his survival. Happy are they, in my opinion, to whom it is given either to do something worth writing about, or to write something worth reading; most happy, of course, those who do both. With his own books and yours, my uncle will be counted among the latter. It is therefore with great pleasure that I take up, or rather take upon myself the task you have set me.

He was at Misenum in his capacity as commander of the fleet on the 24th of August [79 C.E.], when between two and three in the afternoon my mother drew his attention to a cloud of unusual size and appearance. He had had a sunbath, then a cold bath, and was reclining after dinner with his books. He called for his shoes and climbed up to where he could get the best view of the phenomenon. The cloud was rising from a mountain—at such a distance we couldn't tell which, but afterward learned that it was Vesuvius. I can best describe its shape by likening it to a pine tree [today we would compare it to a "mushroom cloud"]. It rose into the sky on a very long "trunk" from which spread some "branches." I imagine it had been raised by a sudden blast, which then weakened, leaving the cloud unsupported so that its own weight caused it to spread sideways. Some of the cloud was white, in other parts there were dark patches of dirt and ash. The sight of it made the scientist in my uncle determined to see it from closer at hand. [This style of explosive mushroom cloud of ash and pumice is now called a "Plinian eruption" in his honor.]

He ordered a boat made ready. He offered me the opportunity of going along, but I preferred to study—he himself happened to have set me a writing exercise. As he was leaving the house he was brought a letter from Tascius's wife Rectina, who was terrified by the looming danger. Her villa lay at the foot of Vesuvius, and there was no way out except by boat. She begged him to get her away. He changed his plans. The expedition that started out as a quest for knowledge now called for courage. He launched the quadriremes and embarked himself, a source of aid for more people than just Rectina, for that delightful shore was a populous one. He hurried to a place from which others were fleeing, and held his course directly into danger. Was he afraid? It seems not, as he kept up a continuous observation of the various movements and shapes of that evil cloud, dictating what he saw.

Ash was falling onto the ships now, darker and denser the closer they went. Now it was bits of pumice, and rocks that were blackened and burned and shattered by the fire. Now the sea is shoal; debris from the mountain blocks the shore. He paused for a moment wondering whether to turn back as the helmsman urged him. "Fortune helps the brave," he said. "Head for Pomponianus."

At Stabiae, on the other side of the bay formed by the gradually curving shore, Pomponianus had loaded up his ships even before the danger arrived, though it was visible and indeed extremely close, once it intensified. He planned to put out as soon as the contrary wind let up. That very wind carried my uncle right in, and he embraced the frightened man and gave him comfort and courage. In order to lessen the other's fear by showing his own unconcern he asked to be taken to the baths. He bathed and dined, carefree or at least appearing so (which is equally impressive). Meanwhile, broad sheets of flame were lighting up many parts of Vesuvius; their light and brightness were the more vivid for the darkness of the night. To alleviate people's fears my uncle claimed that the flames came from the deserted homes of farmers who had left in a panic with the hearth fires still alight. Then he rested, and gave every indication of actually sleeping; people who passed by his door heard his snores, which were rather resonant since he was a heavy man. The ground outside his room rose so high with the mixture of ash and stones that if he had spent any more time there escape would have been impossible. He got up and came out, restoring himself to Pomponianus and the others who had been unable to sleep. They discussed what to do, whether to remain under cover or to try the open air. The buildings were being rocked by a series of strong tremors, and appeared to have come loose from their foundations and to be sliding this way and that. Outside, however, there was danger from the rocks that were coming down, light and fire-consumed as these bits of pumice were. Weighing the relative dangers they chose the outdoors; in my uncle's case it was a rational decision, others just chose the alternative that frightened them the least.

They tied pillows on top of their heads as protection against the shower of rock. It was daylight now elsewhere in the world, but there the darkness was darker and thicker than any night. But they had torches and other lights. They decided to go down to the shore, to see from close up if anything was possible by sea. But it remained as rough and uncooperative as before. Resting in the shade of a sail he drank once or twice from the cold water he had asked for.

> Then came a smell of sulfur, announcing the flames, and the flames themselves, sending others into flight but reviving him. Supported by two small slaves he stood up, and immediately collapsed. As I understand it, his breathing was obstructed by the dust-laden air, and his innards, which were never strong and often blocked or upset, simply shut down. When daylight came again two days after he died, his body was found untouched, unharmed, in the clothing that he had had on. He looked more asleep than dead.

In a second letter to Tacitus a few days later, Pliny wrote:

> By now it was dawn, but the light was still dim and faint. The buildings 'round us were already tottering, and the open space we were in was too small for us not to be in real and imminent danger if the house collapsed. This finally decided us to leave the town. We were followed by a panic-stricken mob of people wanting to act on someone else's decision in preference to their own (a point at which fear looks like prudence), who hurried us on our way by pressing hard behind in a dense crowd. Once beyond the buildings we stopped, and there we had some extraordinary experiences which thoroughly alarmed us. The carriages we had ordered to be brought out began to run in different directions though the ground was quite level, and would not remain stationary even when wedged with stones. We also saw the sea sucked away and apparently forced back by the earthquake: at any rate it receded from the shore so that quantities of sea creatures were left stranded on dry sand. On the landward side a fearful black cloud was rent by forked and quivering bursts of flame, and parted to reveal great tongues of fire, like flashes of lightning magnified in size.
>
> At this point my uncle's friend from Spain spoke up still more urgently: "If your brother, if your uncle is still alive, he will want you both to be saved; if he is dead, he would want you to survive him—why put off your escape?" We replied that we would not think of considering our own safety as long as we were uncertain of his. Without waiting any longer, our friend rushed off and hurried out of danger as fast as he could.
>
> Soon afterward the cloud sank down to earth and covered the sea; it had already blotted out Capri and hidden the promontory of Misenum from sight. Then my mother implored, entreated, and commanded me to escape the best I could—a young man might escape, whereas she was old and slow and could

die in peace as long as she had not been the cause of my death too. I refused to save myself without her, and grasping her hand forced her to quicken her pace. She gave in reluctantly, blaming herself for delaying me. Ashes were already falling, not as yet very thickly. I looked round: a dense black cloud was coming up behind us, spreading over the earth like a flood. "Let us leave the road while we can still see," I said, "or we shall be knocked down and trampled underfoot in the dark by the crowd behind." We had scarcely sat down to rest when darkness fell, not the dark of a moonless or cloudy night, but as if the lamp had been put out in a closed room. You could hear the shrieks of women, the wailing of infants, and the shouting of men; some were calling their parents, others their children or their wives, trying to recognize them by their voices. People bewailed their own fate or that of their relatives, and there were some who prayed for death in their terror of dying. Many besought the aid of the gods, but still more imagined there were no gods left, and that the universe was plunged into eternal darkness for evermore. There were people, too, who added to the real perils by inventing fictitious dangers: some reported that part of Misenum had collapsed or another part was on fire, and though their tales were false they found others to believe them. A gleam of light returned, but we took this to be a warning of the approaching flames rather than daylight. However, the flames remained some distance off; then darkness came on once more and ashes began to fall again, this time in heavy showers. We rose from time to time and shook them off, otherwise we should have been buried and crushed beneath their weight. I could boast that not a groan or cry of fear escaped me in these perils, had I not derived some poor consolation in my mortal lot from the belief that the whole world was dying with me and I with it.

At last the darkness thinned and dispersed into smoke or cloud; then there was genuine daylight, and the sun actually shone out, but yellowish as it is during an eclipse. We were terrified to see everything changed, buried deep in ashes like snowdrifts. We returned to Misenum where we attended to our physical needs as best we could, and then spent an anxious night alternating between hope and fear. Fear predominated, for the earthquakes went on, and several hysterical individuals made their own and other people's calamities seem ludicrous in comparison with their frightful predictions. But even then, in spite of the dangers we had been through, and were still expecting, my mother and I had still no intention o fleaving until we had news of my uncle.

THE AFTERMATH

The sixth and largest surge of ash trapped their ship in the harbor, so Pliny the Elder had no chance of escape. Eventually, boats returned to Pompeii, and found Pliny dead on the docks, apparently asphyxiated by ash or volcanic dust. He was one of thousands of victims; only a few of the original 20,000 people of Pompeii made it out alive. Pompeii was buried so deeply—by almost 20 meters (66 feet) of ash—that the city was abandoned and eventually forgotten (figure 1.2). It wasn't rediscovered until 1748, when well diggers dug down and found the first traces of it in almost 1,700 years. Since then, it has been almost completely excavated, giving us a remarkable window on life in the Roman Empire. Not only are the houses and natural objects well preserved, but the frescoes on the wall still have their original vibrant colors, the tiled mosaics are complete and undisturbed, and the artifacts not made of flammable material are surprisingly undamaged. Most striking was the discovery of empty cavities filled with air in the ash deposits. When these were filled with plaster and excavated, they turned

Figure 1.2 ▲

The ruins of Pompeii, with Mount Vesuvius in the background. (Courtesy of Wikimedia Commons)

Figure 1.3 ▲
Casts of bodies preserved in volcanic ash at Pompeii. (Courtesy of Wikimedia Commons)

out to be natural molds of the bodies of Romans (and their dogs) who died in the ash, asphyxiated by the gases, curled up in protective fetal positions (figure 1.3). Their bodies had been vaporized, but the cavities remained.

Herculaneum was more difficult to excavate since it was covered in a 23-meter-thick (75-foot-thick) blanket of hard volcanic mudflow deposits. Even though it was found in 1709, and digging started in 1738, only part of it has been exposed. Unlike the big city of Pompeii, Herculaneum was a small coastal resort town of about 5,000 people, with many rich villas, as indicated by the houses, jewelry, and other artifacts unearthed there. As in Pompeii, archeologists found cavities left by bodies that had been vaporized, as well as 300 skeletons in death poses. Most were found near the waterfront, where they had been trying to escape but were killed instantly by volcanic gases before their bodies were buried and vaporized.

After the eruption of 79 C.E. wiped Pompeii and Herculaneum from the map—and later from the memory of the Romans—Vesuvius remained active for decades. There was a big eruption in 203 C.E., and again in 472 C.E., with ash reaching as far as Constantinople. Then it went into a dormant phase until the twentieth century, when an eruption in 1906 produced many large flows that killed over 100 people. It erupted again during World War II in 1944, destroying many villages as well as 88 B-25 Mitchell bombers that were involved in Allied invasion of Italy. In the past 70 years, Vesuvius has been relatively quiet, but its history shows that is it still one of the most active and dangerous volcanoes on Earth. Nevertheless, there are now more than 1 million people living on its slopes and more than 3 million around its base, so the next eruption will be far more deadly and disastrous if it is anything like the event of 79 C.E.

The story of Vesuvius and Pompeii is not that different from other great volcanic eruptions. What sets it apart is the observations made by Pliny the Elder and his nephew of the eruption while it was in progress. Instead of treating the event as an act of divine vengeance, both men approached the eruption in a scientific fashion, eager to describe it as a natural phenomenon. This was consistent with the famous 37 volumes of natural history that Pliny the Elder wrote in his lifetime, one of the first such natural histories ever written. The descriptions by Pliny the Younger of the plume of ash that resembled a Mediterranean pine tree and the subsequent stages of the volcanic eruption are the first detailed, scientifically accurate, nonmythological accounts of a volcanic eruption in history. Thus, the accounts of the two Plinys (one of whom died doing scientific observations) mark the beginning of naturalistic observations of earth processes that we now call geology.

FOR FURTHER READING

Beard, Mary. *The Fires of Vesuvius: Pompeii Lost and Found*. Cambridge, Mass.: Belknap Press of Harvard University, 2010.

Cooley, Alison E., and M. G. L. Cooley. *Pompeii and Herculaneum: A Sourcebook*. New York: Routledge, 2013.

De Carolis, Ernesto, and Givoanni Patricelli. *Vesuvius, A.D. 79: The Destruction of Pompeii*. Malibu, Calif.: J. Paul Getty Museum, 2003.

Pellegrino, Charles R. *Ghosts of Vesuvius: A New Look at the Last Days of Pompeii, How Towers Fall, and Other Strange Connections*. New York: William Morrow, 2004.

Scarth, Alwyn. *Vesuvius: A Biography*. Princeton, N.J.: Princeton University Press, 2009.

NATIVE COPPER

We're teaming up with a major Hollywood studio, and we're making a movie called "Copper." It's set on Mars in the 24th century. By then we've got 27 billion people in the world, copper is the world's most valuable metal because everything runs on electricity, and there's no more burning of hydrocarbons.

—ROBERT FRIEDLAND

THE ICEMAN COMETH

On September 19, 1991, two German tourists hiking in the Austrian Alps were taking a shortcut from the officially marked trail at 3,210 meters (10,530 feet). As they trekked along, they spotted a dark object frozen in the ice that they first thought was trash from a previous hiker. When they approached, they could see it was a human head and torso sticking up out of the ice. The body was so well preserved that at first the hikers, and later the coroner and police, thought it was a recent crime victim or a lost hiker who had met a tragic end. It was indeed a lost hiker—but not a recent fatality. In the local morgue, they looked at the clothes and tools and realized the hikers had stumbled upon the mummified body of an ancient human. Later dating established that he lived about 5,300 years ago. Nicknamed "Ötzi the Ice Man" (figure 2.1) after the Ötz Valley where he was found, this "survivor" and his clothes and tools gave us valuable clues about human cultures during the age when copper tools replaced stone tools.

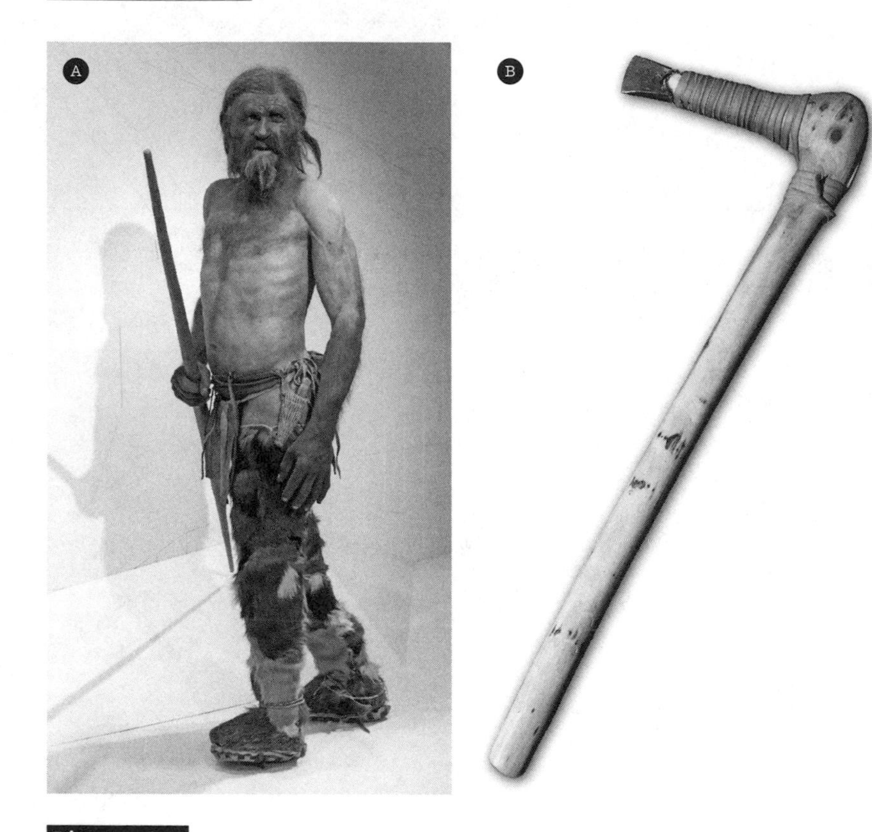

Figure 2.1 ▲
(*A*) "Ötzi the Iceman" and (*B*) the copper-bladed ax he was carrying. (Courtesy of Wikimedia Commons)

At the beginning of human evolution, the first known weapons and tools were made of rocks like flint and obsidian. Known as the Old Stone Age and the New Stone Age (or "Paleolithic" and "Neolithic"), this first step in human tool culture dates back at least 2 million years. But stone tools have their limits—they are brittle and more difficult to shape compared with metal tools. Copper implements represented humanity's first foray out of the Stone Age (giving rise to the Chalcolithic or "Copper Age"), followed by tools and weapons made of a copper-tin alloy, bronze (hence the "Bronze Age"). These metals made tools and weapons that were sharper and lighter in weight than stone tools and that held an edge longer. Armies with metal swords and spearheads had a great advantage and conquered great empires based on the use of metals. Metal tools are also strongly correlated with the sedentary lifestyle of agricultural villages. For one thing, metal made life

Figure 2.2 ▲
Native copper. (Courtesy of Wikimedia Commons)

easier in those communities but required more skill and resources to find. Stone tools can be picked up just about anywhere, but metalworking is a specialized skill that can only be developed in large communities with dedicated metalworkers and with trade networks to obtain the metals.

Unlike most elements and minerals, copper occurs in its pure native element form, as do elemental gold, silver, sulfur, and graphite. Native copper often occurs in huge crystals (figure 2.2). Certain places in the world, such as the Upper Peninsula of Michigan, are famous for their native copper deposits, often relocated across the midwestern United States by glaciers. Some of the earliest copper tools were made of pure native copper, which can be shaped simply by cold hammering. By at least 11,000 years ago there were cultures using implements made from native copper, and copper jewelry in the form of a pendant from the Middle East has been dated to 10,700 years ago. By 7,500 years ago, there is evidence of the next step: an ax-head in Serbia made of copper refined by smelting.

The Copper Age was a transitional period between the Stone Age and the Bronze Age and occurred in different regions at different times: about 4,800 years ago in China, 5,000 years ago in Sumeria and Egypt, 4,280 years ago in northern Europe, and possibly 5,000 or even 8,000 years ago in northern Michigan. Ötzi the Ice Man was carrying what must have been his most prized possession, an ax with a copper head that was 99.7 percent pure copper. He also had lots of arsenic in his hair, suggesting that he was involved in copper smelting during his lifetime. The European Battle-Ax culture dominated many regions from about 7,500 years ago until about 5,300 years ago, when much harder and more durable bronze implements (made by alloying copper and tin) began to be produced in the Middle East, leading to the Bronze Age. But the demand for copper continued, because it is the main component of bronze.

ISLAND OF COPPER

By Greek and Roman times, copper was still widely used, even though the refining of iron and other metals had been developed by then. The Greeks called copper *chalchos*, and mined it from just a few regions in the Mediterranean. The Romans called copper *aes Cyprium*, "metal alloy from Cyprus," because that island was the biggest source of copper at that time. From this we get the Latin word for copper, *cuprum*, which was the name alchemists used later, and the reason that its chemical abbreviation is Cu. Ancient people thought of Cyprus as "the island of copper."

In fact, Cyprus was an important place through all of antiquity, not only because of its strategic location in the eastern Mediterranean, but especially because of its mineral wealth. There is evidence of hunter-gatherer cultures there more than 12,000 years ago, and some of the oldest water wells currently in use in the world, dating to 10,500 years ago, have been found in Cyprus. These ancient peoples apparently caused the extinction of the native Ice Age mammals, such as dwarf hippos and dwarf elephants, which had evolved pygmy forms in isolation on Cyprus, as they did on so many other islands such as Madagascar and Crete. Graves of humans and their pet cats on Cyprus date to 9500 B.C.E., older than any Egyptian mummified cats. The large Neolithic village of Khirokitia dates back to 8800 B.C.E. and is one of the oldest well-preserved Neolithic sites in the world.

Over the next few millennia, Cyprus was conquered and reconquered by nearly every power in the ancient world, primarily to secure its copper supply. The Mycenaean Greeks invaded around 1400 B.C.E., and a second wave of invasion occurred about 1050 B.C.E., when the Greek culture mysteriously collapsed elsewhere. Cyprus was important in Mycenaean and Greek mythology as well. Aphrodite was supposedly born from the foam on one of its beaches, and Adonis had his origin there as well. Cyprus was where the legendary sculptor Pygmalion created his masterwork Galatea, whom the gods turned into a real woman as a reward to the artist. Zeno of Citium, the founder of Stoic philosophy who brought the Stoic ideas to Athens around 300 B.C.E., was a Cypriot.

By the eighth century B.C.E., there were Phoenician colonies on the south coast of Cyprus, importing valuable copper from the mines via their maritime trade empire. The Assyrian Empire conquered the island in 708 B.C.E., then lost it to the Egyptians, who were then replaced by the Persians in 545 B.C.E. During the Ionian revolt of 499 B.C.E., the Cypriots, led by King Onesilus of Salamis, rebelled against the Persian Empire. The revolt was unsuccessful, but the island remained mostly Greek in its culture and largely autonomous. The Greeks on the island welcomed Alexander the Great when he drove out the Persians in 333 B.C.E. Cyprus became part of the Hellenistic Ptolemaic Empire of Egypt when Alexander died and his lands were divided among his generals. Finally, in 58 B.C.E. the Romans conquered Cyprus, and it remained part of the Roman Empire (and its successor, the Byzantine Empire) until Richard the Lionheart, king of England, captured it during the Third Crusade in 1191 and used it as a base for his attempts to win the Holy Land. Richard then sold the island to the Knights Templar, who then sold it to Guy of Lusignan, and eventually it became part of the Holy Roman Empire. However, the maritime city of Venice controlled the island from about 1473 until 1570 C.E., when a full-scale assault by 60,000 Ottoman Turks brought Cyprus under Muslim rule. Even though Cyprus has been ruled by many nations in its history, the bloody warfare it has experienced has largely been based on the tension between its longstanding Greek culture and the more recent Muslim culture that has prevailed since the Ottoman conquest. Finally, in 1974 the island was divided in half, with a Turkish Muslim portion in the northeast (recognized only by Turkey) and the bulk of the island in the southwest being predominantly Greek in culture.

A SLICE OF THE OCEAN

What made Cyprus the richest source of copper known to the ancient world, making it the target of so many battles and invasions? Copper was collected on the island as early as 4000 B.C.E. by scraping up deposits of almost pure copper that were found just lying on the ground. But these deposits didn't last very long, and soon early Cypriots found the original source of this surface copper: the ophiolites of the Troodos Mountains in central Cyprus.

As early as 1813, the pioneering French geologist Alexandre Brongniart coined the word "ophiolite" to describe the peculiar rocks found in the Alps. The word is derived from the Greek *ophis* for "snake," because most of the ophiolitic rocks started out as seafloor lavas made of the black rock called basalt, but were metamorphosed into a rock called serpentine, which gets its name because it resembles smooth, shiny snakeskin. Ophiolites were recognized not only on Cyprus in 1968, but eventually in other places such as Macedonia in Greece and Oman in the Persian Gulf. They always occurred in an odd but predictable sequence of rocks. At the top of the stack were oceanic sediments, which covered lavas that were shaped like blobs and resembled pillows, hence their name—"pillow lavas" (figure 2.3A). At the time, no one knew how these were formed, but today we can see them being created any time lava erupts underwater. If you type "pillow lava eruption" into any search engine, you can find many videos of these dramatic eruptions. As an underwater lava flow moves beneath the ocean, hot new lava breaks through the cracks in the chilled solid outer surface and flows out of the crack like a blob of hot toothpaste (figure 2.3B). It instantly goes from red hot to cold black as the seawater quenches and cools it, leaving the blob or pillow-like shape.

Immediately below the layers of pillow lavas were big vertical walls of congealed lava known as sheeted dikes (figure 2.4). For decades, no one knew what caused them, but eventually geologists realized that they represented the cooled lava fillings of huge vertical fissures in the crust. Magma flowed up through these cracks to feed the pillow lavas, then chilled and cooled in the cracks and formed vertical fins of lava, the sheeted dikes. Below the pillow lavas and sheeted dikes were ancient chilled magma chambers known as layered gabbros, which were the same chemistry and mineralogy as the basalt above. This molten rock had cooled slowly in the magma chamber, rather than erupting as lava, so its crystals were much larger than those found in any volcanic rock. Finally, many ophiolite complexes had rocks known as peridotite in their bottom layers, which we now know to be slices of the upper mantle.

The mysterious ophiolitic rocks of Cyprus and other regions were mapped and described over 150 years ago, but no one could explain how

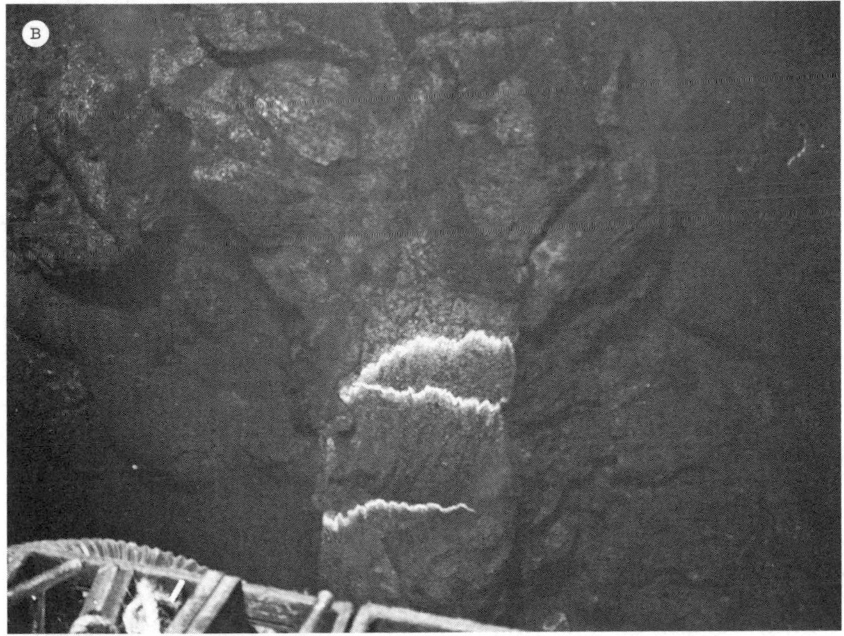

Figure 2.3 ▲

(*A*) Pillow lavas formed by underwater eruptions just west of Port San Luis Pier, California. (*B*) Pillow lavas erupting underwater. ([*A*] Photo by the author; [*B*] Courtesy of Wikimedia Commons)

Figure 2.4 ▲
Sheeted dikes on Cyprus. (Courtesy of Wikimedia Commons)

such curious assemblages of rocks might have formed. The answer did not come until late in the plate tectonic revolution in the late 1960s. Geologists began to realize that when ocean floor spreads apart along a mid-ocean ridge, the natural product of such spreading would be an ophiolite (figure 2.5). The uppermost part of the spreading crust would be made of pillow lavas from magmas that had reached seawater and then chilled. Beneath them would be vertical cracks formed when the crust pulled apart, which filled with lava to make sheeted dikes. Finally, the source of all this volcanic material was the layered gabbro of the magma chamber beneath it, and beneath that some ophiolites even contained slices of the upper mantle.

But how did rocks formed deep on the ocean floor end up on land in places like Cyprus? This too was a result of plate tectonics. When two plates are pushed together, one plate (made of oceanic crust) usually slides beneath the other and plunges down to the mantle in what is known as a subduction zone. The majority of the subducting crustal slab slides smoothly down into the mantle without a fuss, although much of the ocean floor sediment on top of the slab is sliced and scraped up and plastered into the edge of the

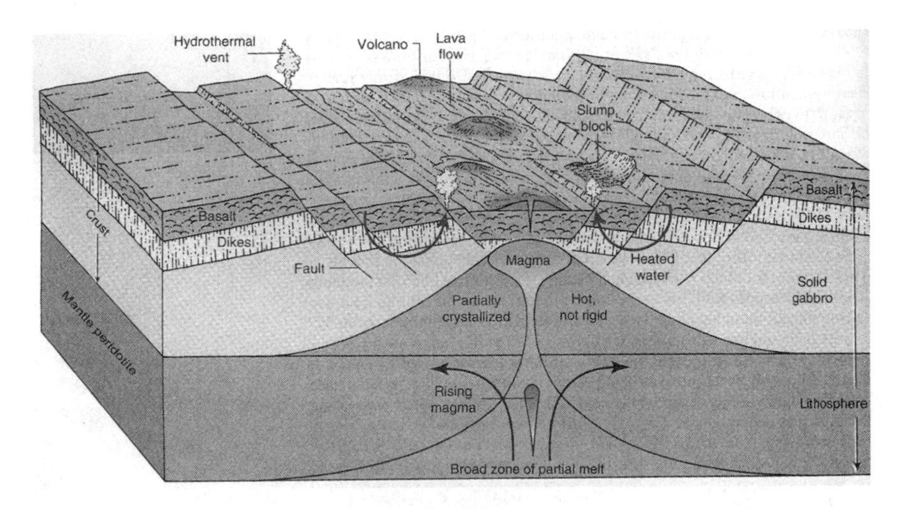

Figure 2.5 ▲
A formation of the ophiolite suite in mid-ocean ridges. (Redrawn from several sources)

overlying plate in what is known as an accretionary wedge. Occasionally, however, a slice of oceanic crust itself gets chopped off the downgoing slab and pushed into the accretionary wedge. This can be seen in California's ophiolites in the western Sierran foothills, in the Klamath Mountains, and particularly in the Pacific Coast Ranges, all formed when California used to be a subduction zone. Ophiolites are also formed when two continents collide and the oceanic crust between them gets trapped and uplifted into a mountain belt, as happened in Cyprus when the African plate collided with the Anatolian plate of Eurasia.

AT THE BOTTOM OF THE SEA

The mid-ocean spreading zone origin of ophiolites was confirmed by many measurements and observations of modern oceanic ridges during the 1970s, but why were ophiolites like those in Cyprus so rich in minerals like copper? This puzzle was solved in 1977, when a dramatic discovery was made on the ocean floor. Using the small research submarine *Alvin*, which could dive almost 4,800 kilometers (14,800 feet) beneath the waves, scientists at Woods Hole Oceanographic Institution spent hours investigating the seafloor on mid-ocean ridges.

At such depths, there is total inky darkness. The temperature of the water is just above freezing, and it is so deep and under the weight of so much seawater that it exerts pressures of 5,800 pounds per square inch, 400 times what we experience in our atmosphere at sea level. This would crush the submarine if it were not specially built for the project. *Alvin*'s scientists saw not only huge areas of pillow lavas, but an even greater surprise: superheated plumes of black, mineral-rich boiling water shooting out of chimneys made of sulfide minerals like pyrite ("fool's gold" or iron sulfide, whose formula is FeS_2) (figure 2.6). Nicknamed "black smokers," these chimneys were formed when cold seawater percolated down through cracks to the hot magma below, then boiled and rose up as hot springs full of sulfide minerals. These minerals included not only pyrite, but copper sulfides (covellite or CuS; chalcocite or Cu_2S; and chalcopyrite or $CuFeS_2$), zinc sulfides, and lead sulfides, plus manganese, silver, gold, and other metallic minerals. These had been dissolved out of the crustal rock by the superheated water percolating through the cracks, and then precipitated when they met cold seawater and crystallized out of the hot solution.

Even more surprising was that these black smokers supported an entire community of organisms that were entirely new to science. There were giant clams over a meter long, long tubeworms, odd-looking albino crabs, and many other creatures that had never been seen before. I vividly remember attending a seminar at Woods Hole in 1978, when I was doing my graduate research there, and seeing scientists present these strange animals to their peers for the first time.

Scientists eventually realized that all these strange creatures represented a unique biological community found only in the deep-sea vents of mid-ocean ridges. Unlike most ecological communities whose food chain is based on plants that convert sunlight into organic matter via photosynthesis, these communities lived in a world with no sunlight. Instead, they were based on a food chain that started with sulfur-reducing bacteria that lived in the superheated sulfur-rich seawater and converted the energy of the hot springs into carbon through chemosynthesis. All the animals higher up in the food chain fed on these bacteria or on the smaller organisms that ate the bacteria. Thus the generalization we learn in science class that all food pyramids begin with plants at the bottom is not true on the seafloor. Instead of photosynthetic communities based on plants, these are chemosynthetic communities based on bacteria.

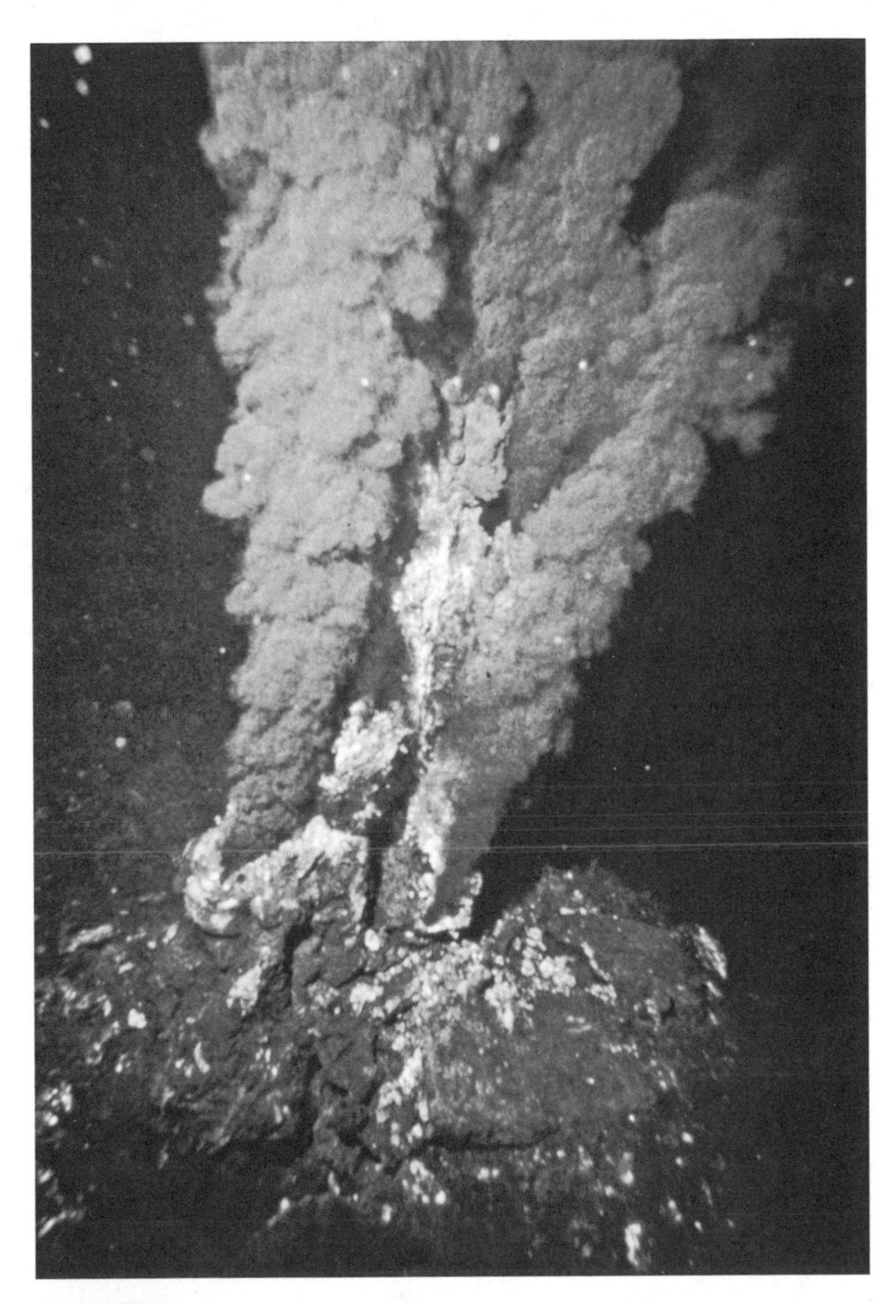

Figure 2.6 ▲

The superheated seawater in mid-ocean ridges creates plumes of boiling water full of dissolved minerals, which precipitate to form chimneys called "black smokers." (Courtesy of Wikimedia Commons)

The black smokers also solved the mystery of the copper-rich ophiolites of Cyprus. Black smokers naturally concentrate minerals such as iron, copper, zinc, lead, manganese, and other metal sulfides: the superheated water dissolves these metals and then precipitates them in the chimneys of the black smokers. Thus the ancient copper miners of Cyprus were unknowingly getting their wealth from a chunk of Jurassic seafloor with ancient black smokers that had been lifted to the tops of the Troodos Mountains by plate tectonics.

FOR FURTHER READING

Fowler, Brenda. *The Iceman: The Life and Times of a Prehistoric Man Found in an Alpine Glacier*. Chicago: University of Chicago Press, 2001.

Lienard, Jean. *Cyprus: The Copper Island*. Paris: Le Bronze Industriel, 1972.

Nicholas, Adolphe. *The Mid-Ocean Ridges: Mountains Below Sea Level*. Berlin: Springer, 1995.

Searle, Rorger. *Mid-Ocean Ridges*. Cambridge: Cambridge University Press, 2013.

03 THE "ISLES OF TIN" AND THE BRONZE AGE

CASSITERITE

Bronze was as valuable a raw material as oil is today.

—ARCHEOLOGIST KRISTIAN KRISTIANSEN

TO THE ENDS OF THE EARTH

In ancient times, travel across the oceans was a perilous business. Ships could not navigate far from the shore; they traveled slowly with their small sails or crews of slaves at the oars; and their maps were poor. Most ancient civilizations were land-based and sent their armies, not navies, into battle. Only a few cultures, such as the Phoenicians in the eastern Mediterranean or the Greeks sailing between their many islands, developed a significant maritime culture. Even seafaring cultures such as the Phoenicians, who had the first decent maps, had no way of determining their east-west longitude and precisely positioning themselves on those maps. Thus, they sailed only short distances across open water and close to the shoreline whenever possible.

The Romans conquered all the lands that surrounded the Mediterranean Sea, and they called it *Mare nostrum*, "our sea," because they ruled all its shores. (The Roman name *Mediterranean* literally means "the sea in the middle of the land," since it was in the middle of their empire.) Nevertheless, they employed limited naval forces in their battles, relying instead on their well-trained legions.

Very few Mediterranean seafarers dared venture into the unknown waters of the Atlantic Ocean, which was the edge of the earth to peoples of that time. Indeed, the name "Atlantic" comes from the old Greek myth of the Titan named Atlas, who supposedly held the heavens or the earth on his shoulders. The Atlas Mountains, which cross the Straits of Gibraltar, take their name from him; another name for the rocks that frame the straits (Gibraltar in Spain and Jebel Musa in North Africa) is the "Pillars of Hercules," reflecting the Greek myth of Hercules briefly taking over the burden from Atlas as part of his twelve labors. Any place beyond those treacherous waters was considered the realm of the unknown. This is part of the reason that Plato located his mythical Atlantis beyond the Pillars of Hercules.

Nevertheless, seafaring nations that had discovered the copper deposits of Cyprus (chapter 2) were desperate for another metal as well: tin. In the ancient world, tin was of enormous importance, because tin mixed with copper (5 to 20 percent tin, the rest copper) makes the alloy known as bronze, which was harder than any other metal then known and easier to shape than either tin or copper. This first metallic alloy could be made into superior tools and weapons and ushered in the Bronze Age.

Tin is such a rare element in the rest of Europe that the demand for it was extraordinary. Many traders traveled long distances to find it, especially after the deposits in the Mediterranean were exhausted. The Phoenicians were the first reach the tin deposits of southwest Britain, and they kept the source of their tin wealth a closely guarded trade secret. According to legend, a ship captain from Carthage (the Phoenician city-state located in what is now Tunisia) would wreck his vessel rather than risk being followed by Greek, or later Roman, ships to find the secret source of this precious metal. The Greeks, battling the Phoenicians for control of the Mediterranean, knew of the legendary "isles of Tin" they called the "Cassiterides," but not their location (figure 3.1). As many maps show, early sailors thought the Cassiterides were islands, not part of the land they later discovered to the east and called "Brittania." This is the first mention of the British Isles recorded in ancient history—the isles of Tin.

Nevertheless, the legendary sources of the Phoenician tin were widely discussed by ancient authors. As early as 500 B.C.E. Hecataeus of Miletus wrote about lands beyond Gaul where tin was obtained. About 325 B.C.E. Pytheas of Massalia sailed to Britain, where he found a flourishing tin trade, according

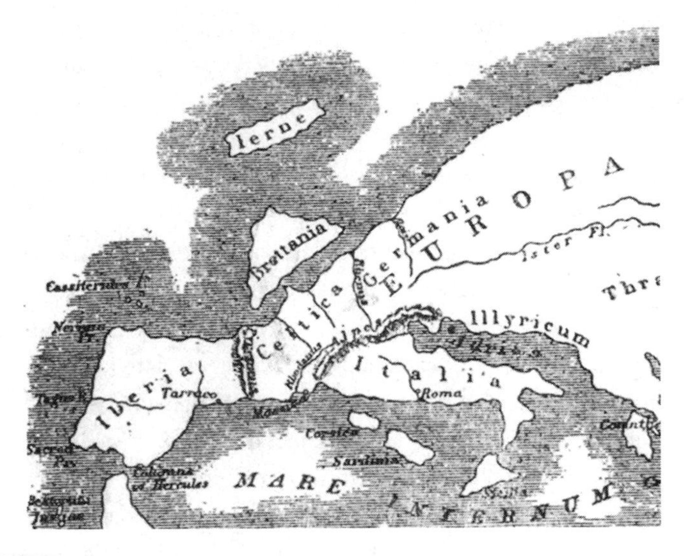

Figure 3.1 ▲
An ancient map of the world from Strabo's geography (written around 23 B.C.E.) showing the Cassiterides as "islands" separate from "Brettania," which was how the ancients perceived Cornwall. (Courtesy of Wikimedia Commons)

to later reports of his voyage. The Greek astronomer and geographer Posidonius referred to the tin trade with Britain around 90 B.C.E.

Many later scholars speculated on what region was the actual Cassiterides. Some identified the Cassiterides with the Scilly Isles and argued that this showed that the Phoenicians traded to Britain. However, there is no tin mining on the Scilly Isles, except for minor exploratory excavations. In more recent times, it has become clear that the "islands of Cassiterides" were indeed the Cornish Peninsula on the western tip of Britain, and not islands at all.

Ancient tin mining in Britain was described by Diodorus Siculus around 1 B.C.E.:

> They that inhabit the British promontory of Belerion by reason of their converse with strangers are more civilised and courteous to strangers than the rest are. These are the people that prepare the tin, which with a great deal of care and labour, they dig out of the ground, and that being done the metal is mixed with some veins of earth out of which they melt the metal and refine it.

> Then they cast it into regular blocks and carry it to a certain island near at hand called Ictis for at low tide, all being dry between there and the island, tin in large quantities is brought over in carts.

Tin had already been mined in southwestern Britain (Devon and Cornwall) as early as 2100 B.C.E., mostly by local Celtic tribesmen. They traded with Phoenicians from places that are now known as Lebanon and Syria. Later, the tin trade from Cornwall was handled by the Veneti, the Breton people of Brittany in northwestern France, who shared the ancient Cornish language and culture. Southwestern Great Britain was never glaciated, so the ore deposits are near the surface and not bulldozed away by glacial ice or covered by a blanket of glacial debris. The ore even occurred in the gravel of streams, which were the first deposits to be exploited by prehistoric people, but eventually the miners began to dig short trenches directly into the ore bodies.

Tin mining continued through ancient times, the Middle Ages, and right up through the twentieth century. It was so important that in 1305, King Edward I created the Stannary Courts and Parliaments to administer and regulate the tin mines. They were the primary forms of government in Devon and Cornwall for centuries. The tin miners scraped up the shallow surface deposits of tin from deep trenches and then processed them through a stamp mill run by a waterwheel. The crushed ore was then melted in a smelter and poured into molds made of granite to make rough ingots of tin. These were taken to the Stannary Courts for weighing and assaying.

The wealth of the tin mines made them a tempting target, and in 1497 King Henry VII raised the taxes on the tin mines to pay for his wars in Scotland, in violation of the previous rules set up by Stannary Courts. The Cornish miners were so angry that they revolted against the king, and their armies moved across southern England virtually unopposed for weeks. They roamed north to Bristol and reached as far as Kent in southeastern England before retreating. Finally, on June 17, 1497, the Cornish rebels faced an army of 25,000 troops supporting the king at the battle of Deptford Bridge. The bridge and the battleground are long gone, covered up by the urban sprawl of southeastern London. The rebel army was much smaller, with no cavalry or artillery, so they were at a decided disadvantage. The Cornish archers held off the royal forces trying to cross the bridge for some time, but without reinforcements nearby, they were soon overwhelmed, and the rebel army was cut to pieces, losing as many as 2,000 men.

CORNISH TIN MINERS

This battle was the last time the Cornish fought the Crown in an open rebellion, but the people of Cornwall are a proud and tough race with their own distinct dialect and local cultures. They have always considered themselves distinct from the rest of Britain, and they fly the black flag with the white cross with pride all over Cornwall. In later years, the many isolated coves and towns in Cornwall made it an excellent place for smugglers trying to evade royal taxes and tariffs, so there were many rough seafarers operating in the region. They were the inspiration for the *Pirates of Penzance*, the famous Gilbert and Sullivan operetta.

In the 1700s and 1800s, tin was in huge demand not for bronze weapons, but for pewter (85 to 90 percent tin alloyed with copper, antimony, or lead) for bowls, cups, plates, utensils, and vessels. The biggest demand, however, was for its use in the original "tin can" and "tinfoil," which were essential to sealing food in airtight containers. In fact, it could be said that tin changed the course of modern warfare and made the great empires of the 1700s and 1800s possible. The tin container made it feasible for armies and navies to supply themselves during long voyages and campaigns; obtaining sufficient food had always been a problem for military logistics. Tin cans were actually invented at the request of Napoleon Bonaparte himself. As Napoleon famously said, "An army marches on its stomach."

Tin was used in many other types of implements and objects, since it was an easy metal to melt, mold, and work, and did not corrode. Generations of European boys played with "tin soldiers" like the one in Hans Christian Andersen's story "The Steadfast Tin Soldier." Before aluminum foil became common, actual tinfoil was the main type of foil used to seal containers and it was also used in electrical devices; many people still call aluminum foil "tinfoil" for that reason. During this time, "Cornish tin miner" became a common meme in British culture, comparable to "Welsh collier." Tin continues to be used in limited quantities, especially as a noncorrosive liner for steel containers and in solder for electronics. For example, the Apple iPad uses 1 to 3 grams of tin, and in just two of its components there are 7,000 solder points. Tin is still the most important metal in most electronics produced today.

Nevertheless, the replacement of tin cans and tinfoil with aluminum made tin less important in the modern world. The Cornish deposits became too depleted to be worth mining in the face of a declining price and market, so the mines began to close down, starting in the 1920s. In addition, enormous tin deposits were found in Peru and Bolivia, making this region the tin-mining capital of the world for much of the twentieth century. Then significant deposits were found in China, Australia, Malaysia, and other countries. Huge deposits were found in the Congo, leading to their exploitation by various violent rebel groups fighting their governments in Africa. Currently, most of the world's tin comes from Indonesia, China, and Malaysia.

WHERE DID THE TIN COME FROM?

The area around Devon and Cornwall was the main source of tin in Europe. The tin is directly due to the intrusion of Carboniferous and Permian granitic rocks (such as at Dartmoor and at Land's End). This took place during the Variscan Orogeny, when portions of a microplate known as Armorica collided with southern England, crumpling its rocks and then intruding it with magmas melting from the root of the mountain range.

The intrusions brought many veins of mineral-rich magma up near the surface. They also superheated the groundwater, creating hydrothermal deposits where groundwater percolated through the surrounding Devonian bedrock and scavenged rare elements and then precipitated into veins rich in those minerals. The veins are rich in copper, lead, zinc, and silver, but this area is most famous as one of the world's largest deposits of tin in the form of the mineral *cassiterite* (tin oxide, or SnO_2), which gets its name from the old legend of the Cassiterides. Cassiterite is a striking, metallic, silvery-gray mineral that can form twinned crystals and occasionally even forms crystals shaped like bipyramids, two pyramids joined together end to end (figure 3.2).

Even though the tin mines of Cornwall are mostly closed, it is possible to visit them today and get a glimpse of what was once a proud and powerful industry that helped propel Britain's industrial might. A historic tin mine open for tours is Geevor Tin Mine in Cornwall, which is just north of Land's End, the westernmost point in Britain. Operational between 1840 and 1990, it was one of the largest tin mines in Britain during its heyday, producing over 50,000 tons of tin over its lifetime. During the 1880s, it had almost 200 workers in this one mine alone; at its peak, it employed 270 people. Today it is a living history site recognized by UNESCO as a World Heritage Site, with guided tours of the former tin-mining operations by men and the sons of men who once worked there.

Figure 3.2 ▲
Crystals of cassiterite—or tin oxide—in its twinned crystal form. (Courtesy of Wikimedia Commons)

Tin mining was always a dangerous, dirty occupation. Even the surface mining of tin led to disputes over timber, water, and pastures. All of the underground operations were conducted in narrow mine shafts that plunged deep into the earth following the mineral veins. At the surface, the most prominent landmark at Geevor Tin Mine is the headframe (figure 3.3), which held the machinery and the spool of cables to raise the elevators that lowered the cages full of men and equipment below ground and then raised their rolling mine cars full of ore to the surface as they worked. The tin mining was originally done with pickaxes and shovels digging into the hard bedrock, but eventually hydraulic drills and hammers were used to drill holes in the rock to place their dynamite in (figure 3.4). For decades, this drilling created a huge amount of rock dust, and miners lived short lives because their lungs were scarred by silicosis. The eventual adoption of water-cooled drills got rid of the rock dust problem and lowered the death rate of the miners.

At the end of the day, miners would load charges of dynamite into the drill holes, then leave the mineshaft and blow the dynamite. After the dust had settled overnight, they would return to their active operation the next morning and load all the ore and waste rock into small mining carts to be

Figure 3.3 ▲

The headframe at Geevor Tin Mine Museum, Cornwall. (Photo by the author)

Figure 3.4 ▲
Underground tin mining. (Courtesy of Wikimedia Commons)

transported to the surface. They had to excavate many horizontal tunnels (adits) to reach ore veins or drain out water, and they connected those tunnels with vertical shafts that helped them reach veins at deeper levels and acted as drainage. Multiple short horizontal tunnels from the main mineshaft would be excavated to reach an ore-bearing vein in a technique known as stoping. Over the years, the Geevor Tin Mine reached deeper and deeper veins, some deep below the Atlantic Ocean and many miles from the mine opening. Eventually, there were almost 85 miles of tunnels, many of which followed the tin veins deep under the floor of the sea and required nonstop pumping to get rid of seawater. Other mines, such as Dolcoath, were among the deepest mines in the world in the 1920s, reaching 1,067 meters (3,500 feet) below the surface.

Once the carts full of rock were brought to the surface, the raw ore was dumped into a stamp mill, which crushed the rock down to fine sand-sized particles. Then the challenge was to separate the valuable ore from the worthless waste rock. Most of the time the raw material was flushed through a water system, which separated the denser metallic minerals from the lighter waste rock. Older mines had a series of large water-filled pits called buddle pits, which were lined with clay and were used to let the dense metals separate out from the lighter minerals.

Huge amounts of ore rock had to be processed to get at the valuable minerals in them. In the Cornish mines, the ore grade was typically 1 percent or less, so that 100 tons of rock would have to be mined to produce 1 ton of pure tin.

More modern tin mines like Geevor had a building with hundreds of magnetic separators (figure 3.5) that ran the powdered ore on a water table covered by riffles and used a magnetic field beneath the top of the table to hold back the magnetic minerals in the ore while the nonmagnetic minerals washed down toward the waste-collection area. Finally, the concentrated metallic residue was collected and sent to the smelter, where it was melted and separated into different metals (tin, copper, zinc), then cast into ingots that could be sent to the market. One of the weaknesses of the Cornish mining region was the absence of local coal or other fuels to heat their smelters, so they had to ship the processed ore to other regions for final smelting.

Figure 3.5 ▲

The magnetic separator at Geevor Tin Mine. The riffles help separate the heavy tin and other metals from the lighter waste rock. Then magnets beneath the table concentrated the metallic minerals at the top of the table, and let the waste material flow out the bottom. (Photo by the author)

COLLAPSE OF THE TIN EMPIRE

The Cornish tin mines were becoming less and less profitable by the early twentieth century, when richer deposits that were more economical to work were found in Bolivia and Southeast Asia. Nevertheless, tin mining continued in Cornwall through much of the twentieth century. The price of tin was partially supported by a worldwide cartel, the International Tin Council, which internally regulated the production of tin around the world to keep it scarce and boost its price in the commodities market. When necessary, they bought surplus stocks from Cornwall or Malaysia to maintain the price.

Nevertheless, the demand for tin was steadily declining, with cheap aluminum replacing it in many items. Mining in underdeveloped countries undercut the prices for the more expensively produced Cornish tin. Eventually, the International Tin Council ran out of the money needed to keep buying up surplus tin to support the price, and in October 1985 the cartel suddenly collapsed, despite several efforts to revive it. Now the price of tin is much lower, determined by the cheap tin mined in low-wage countries and limited demand. This has come at an increased hazard to miners, especially in Indonesia, where the deposits are getting deeper and more hazardous to work. There, the miners, still using pickaxes and shovels, work for a mere $5 a day, and there are no safety regulations, such as terracing the open-pit deposits, to prevent cave-ins. In 2011, about one miner a week was killed in these mines, but there is no impetus to regulate the mines or improve their safety.

The collapse of the world tin cartel in 1986 marked the end of tin mining in Cornwall and Devon. The mine in Devon that most recently produced tin ore was Hemerdon Mine near Plympton in the 1980s. The last Cornish tin mine in production at South Crofty closed in 1998. When you tour the mines like Geevor, the final and most poignant stop is the miners' locker room, where many of them left their all their gear and mining clothes behind after their last shift when the entire mine was shut down without warning in 1986. The miners and tour guides still grumble about how the rich investors, owners, and commodity traders of the cartel made a profit from insider knowledge before the sudden collapse of prices, but the miners got no warning and had no futures after the mines closed. A graffito outside one Cornish mine written circa 1999 reads: "Cornish lads are fishermen and Cornish lads are miners too. But when the fish and tin are gone, what are the Cornish boys to do?"

Touring the haunted hills of Cornwall and seeing all the long-abandoned mining operations is a sobering experience. It is the final stage in the long history of the metal that ushered in the Bronze Age, powered the Industrial Revolution, made it possible to feed large armies and navies food preserved in tin cans, and still is one of most important metals in modern electronics. We may no longer be in the Bronze Age, but tin is still important, even in the Computer Age.

FOR FURTHER READING

Atkinson, R. L. *Tin and Tin Mining*. London: Shire Library, 2010.

Price, T. Douglas. *Europe Before Rome: A Site-by-Site Tour of the Stone, Bronze, and Iron Ages*. Oxford: Oxford University Press, 2013.

04 | "NO VESTIGE OF A BEGINNING": THE IMMENSITY OF GEOLOGIC TIME

ANGULAR UNCONFORMITY

The mind seemed to grow giddy by looking so far into the abyss of time.

—JOHN PLAYFAIR

IN THE BEGINNING . . .

For almost 2,000 years, nearly all scholars followed a literal interpretation of the Bible as their guide for the origin and history of the earth. Even as late as the mid-1700s, natural historians still thought the earth was only a few thousand years old, perfect and unchanging except for erosion and decay brought about by Adam's sin. According to Church dogma, the world otherwise would have been perfect and unchanging. The thinking of prominent natural historian John Woodward (1665–1728) was typical of his time when he wrote in 1695: "The terraqueous globe is to this day nearly in the same condition that the Universal Deluge left it; being also like to continue so till the time of its final ruin and dissolution, preserved to the same End for which 'twas first formed."

Ideas about the age of the earth were dictated by biblical dogma. For example, in 1654 James Ussher (1581–1656), the Anglican archbishop of Armagh, Ireland (which was mostly Catholic at the time, so he had few Anglicans to minister to), used the ages of the patriarchs in the Bible to calculate that Creation happened on October 23, 4004 B.C.E. Another scholar, John Lightfoot, placed the time at 9 A.M. (Neither of them explained how days and nights could be recognized before there was a sun or Earth.)

Of course, the Bible doesn't give consistent accounts of how much time lapsed between Creation and Noah's flood, let alone the time afterward, so a lot of guesswork was involved. Nevertheless, Ussher's estimate was the pinnacle of scholarship for its time, incorporating what was known of the history of the Hebrews, Babylonians, Persians, Greeks, and Romans, so we must respect this estimate for the honest attempt that it was—even though we now know it was about a million times too short.

THE ENLIGHTENMENT

For more than a century, the power of the Church over European scholarship meant that this estimate was not challenged. However, during the Enlightenment, the hold of religious dogma over scholars and scientists began to weaken. Some, like the Frenchman Georges-Louis Leclerc, count Buffon (1707–1788) in 1779, suggested the earth was as old as 75,000 years, at least 10 times the estimate based on biblical chronology.

In the second half of the eighteenth century, scholars and natural historians began to question the power of both the Church and the aristocrats. They used rationality, evidence, and critical thinking to challenge the powers that be and the way things were. They focused on examining the sources of human knowledge, the justification for the power of governments and religious leaders, and the unquestioned assumptions of past centuries. In France, the Enlightenment was based in the salons and culminated in the great *Encyclopédie* edited by Denis Diderot with contributions by hundreds of leading intellectuals such as Voltaire (1694–1778), Jean-Jacques Rousseau (1712–1778), and Montesquieu (1689–1755). The Enlightenment in England was first inspired by Isaac Newton's (1643–1727) transformation of physics and our understanding of the universe. It was also led by John Locke (1632–1704), whose ideas about government and religion were an inspiration to Enlightenment-oriented Americans such as Thomas Jefferson (1743–1826), Benjamin Franklin (1706–1790), and other Founding Fathers. Among these was Thomas Paine, who wrote polemics not only against English rule in the Americas but against religion and the Bible as well. Immanuel Kant (1724–1804) revolutionized the field of philosophy in the German-speaking world, and Gottfried Leibniz (1646–1716) made many advances in science and mathematics, especially the invention of calculus (a slightly different version of which was created by Isaac Newton).

To the surprise of many people, Edinburgh was a major intellectual center and the hub (along with Glasgow) of the Scottish Enlightenment. Nicknamed the "Athens of the North," Edinburgh featured many neoclassical buildings and had a reputation for learning, like that of its ancient namesake. In Tobias Smollett's novel *The Expedition of Humphry Clinker* (1771), one character referred to Edinburgh as a "hotbed of genius," and historian James Buchan describes it well in the title of his book *Crowded with Genius*.

Why was such a small provincial city like Edinburgh one of the intellectual centers of the world, surpassing even great cities like London and Paris? As Arthur Herman pointed out in his book *How the Scots Invented the Modern World*, a number of factors contributed to this ideal environment for free thought and intellectual ferment. The first was the political stability and economic boom that came after union with England in 1707. Scottish traders had become rich with their trans-Atlantic trade (especially in tobacco), and their wealth endowed many institutions, especially universities. Except for the troubles with the Catholic Jacobites and Bonnie Prince Charlie that ended after the Battle of Culloden in 1745, Edinburgh experienced political stability and peace for most of the 1700s. After 1745, the Scots tried hard to emulate the English and to succeed in English society and culture.

A second important influence was the religious climate of the city and the absence of religious persecution. After a young Scotsman, Thomas Aikenhead, was hanged in 1697 for blasphemy, the power of religious leaders in Edinburgh rapidly began to wane. Part of the reason was that Scotland was split between Catholics (especially Scottish royalty and the Highlanders) and the mostly Presbyterian followers of John Knox, who were influenced by the Calvinists, with a minority of Anglicans following England's official church (the Lowland Scots). This is in contrast to England, where anyone who was not a member of the Anglican Church had little chance of advancement, or France, where the Catholic Church was powerful and the aristocracy was corrupt.

The Presbyterians were great believers in every man reading the Bible for himself, and they instituted public schools throughout Scotland, so that by the late 1700s Scotland had the highest literacy rate in the world. By that time, Scotland had five major universities, while England had only two. There were many newspaper and book publishers. In Scottish intellectual life, culture was oriented toward books. In 1763 Edinburgh had six printing houses and three paper mills, but by 1783 there were 16 printing houses and 12 paper mills. Thus, Edinburgh became a major center for the English-language book trade.

Intellectual life revolved around a series of social clubs that began to spring up in Edinburgh in the 1710s. One of the first and most important was the Political Economy Club, aimed at creating links between academics and merchants. Other clubs in Edinburgh included the Select Society, founded by artist Allan Ramsay (1713–1784), philosopher David Hume (1711–1776), and economist Adam Smith (1723–1790). A later organization was the Poker Club, formed in 1762 and named by the historian and philosopher Adam Ferguson (1723–1816) for its aim to "poke up" opinion on many public issues.

Historian Jonathan Israel points out that by 1750 almost all the major cities in Scotland had an intellectual infrastructure of mutually supporting institutions, such as universities, reading societies, libraries, periodicals, museums, and Masonic lodges. The Scottish network was "predominantly liberal Calvinist, Newtonian, and 'design' oriented in character which played a major role in the further development of the transatlantic Enlightenment." Bruce Lenman says their "central achievement was a new capacity to recognize and interpret social patterns."

Some of the greatest advances came in philosophy, where the freedom to think and question and argue without religious restraint led to great breakthroughs. Most of the great late eighteenth-century Scottish Enlightenment figures were influenced by the legendary Francis Hutcheson (1694–1746), who was professor of moral philosophy at the University of Glasgow from 1729 to 1746. His ideas inspired many later philosophers, such as Adam Smith, David Hume, Immanuel Kant, and Jeremy Bentham, all of whom emphasized the practical, utilitarian, and realist threads of philosophy, versus the more abstract ideas of the philosophers who preceded them.

JAMES HUTTON

Among the geniuses of the Scottish Enlightenment was the legendary skeptical philosopher David Hume; the economist Adam Smith, whose work *The Wealth of Nations* first described capitalism; the chemist Joseph Black (1728–1799); James Watt (1736–1819), the co-inventor of the modern steam engine that powered the Industrial Revolution; and a quiet young gentleman named James Hutton (figure 4.1). Hutton was born in Edinburgh on June 3, 1726, and was the son of a prominent merchant and city officeholder.

Figure 4.1 ▲
Portrait of James Hutton. (Courtesy of Wikimedia Commons)

Though Hutton's father died when his son was quite young, James managed to get an education in the local grammar school and at the University of Edinburgh. Even though he was primarily interested in chemistry, he entered the legal profession. But as a lawyer's apprentice, he spent a lot of time entertaining his fellow clerks with chemical experiments rather than copying legal documents. He, along with his friend James Davie, was also deeply interested in investigating the manufacture of sal ammoniac (now known as ammonium chloride, NH_4Cl) from coal soot. As a result, Hutton was released from his law apprenticeship before his first year was out, and he turned to the study of medicine, as it was the only way at that time to study chemistry or any other kind of natural science. He spent 3 years at the University of Edinburgh, then two in Paris, and finally was granted a medical degree in Holland in September 1749. (His departure to Paris also allowed him to escape scandal back home in Scotland for fathering an illegitimate child.)

But the actual practice of medicine held no appeal for Hutton. His association with Davie in developing an inexpensive method for the manufacture of sal ammoniac proved successful and profitable, and this business then allowed Hutton to take time to manage his family's farms, especially the one known as Slighhouses in Berwickshire, Scotland. He used his training to experiment with the latest techniques in agriculture with great success. Clearing the landscape, cutting ditches, and draining his farm created many fresh cuts into the local bedrock, which fascinated him. In 1753 he wrote that he had "become very fond of studying the surface of the earth, and was looking with anxious curiosity into every pit or ditch or bed of a river that fell in his way." By 1765 the farms and the company producing sal ammoniac were prospering, and he had enough income to leave the work to tenant farmers in 1768. He then moved back to Edinburgh to follow his scientific interests.

Hutton had inherited his father's properties and had income from the farms and the sal ammoniac business, so he had no need to work for a living. As a result, he had plenty of free time to socialize with his friends, particularly Adam Smith and Joseph Black. Together, they formed another discussion group, the Oyster Club, which met every Friday afternoon at two but at a different tavern each week, since the meetings were often a bit too popular. There they would convene to discuss art, architecture, philosophy, politics, physical science, and economics, each giving a brief update on his special project. The discussions were, in Hutton's words "informal and amusing despite their great learning." Other members included James Watt and John Playfair, a mathematician who became Hutton's main disciple among geologists, and many other scholars and natural philosophers from the university. When he visited Edinburgh, Benjamin Franklin was treated as an honored guest. A Swiss chemist who visited Edinburgh described the Oyster Club in these words: "We have a club here which consists of nothing but philosophers. Drs. Adam Smith, Hutton, Cullen, Black, Mr. McGowan belong to it, and I am also a member of it. Thus I spend once a week in the most enlightened and agreeable, cheerful and social company."

UNIFORMITARIANISM

As a gentleman-farmer and landowner maintaining and improving his family's farms in southeast Scotland, Hutton had studied how soils form, how sediments erode, and how layers of sediment are deposited as they flow down rivers and into the sea. From this, Hutton gained insights about rock weathering, and how slowly sediments are formed and deposited. He visited the ancient Roman fortification near the Scottish-English border known as Hadrian's Wall (figure 4.2) and noticed that it had not weathered or broken down much in the more than 1,600 years since it was built in 122 C.E. From this, Hutton realized that the process of weathering down whole mountain ranges would take much longer.

Hutton devoted his time to extensive scientific reading and traveled widely to inspect rocks and observe the actions of natural processes. Using the basic principle of natural law that all Enlightenment scholars followed, Hutton applied the principle of naturalism to the earth as well. In his mind, supernatural catastrophes ("catastrophism") such as the biblical deluge of Noah were useless as scientific explanations, because they could not be subjected to test and examination by natural principles or evidence. Instead, Hutton argued that natural laws and processes that operate today must have operated the same way in the past. This is often called the principle of uniformitarianism, or in geologist Archibald Geikie's words, "the present is the key to the past." Hutton's ideas, including the uniformitarian principle, were first formally presented to the Royal Society of Edinburgh in 1785. Two of his papers were published in 1788 in the *Transactions of the Royal Society of Edinburgh* under the title "Theory of the Earth; or an

Figure 4.2 ▲

When Hutton visited it in the 1770s, Hadrian's Wall had not eroded or changed significantly in the more than 1,600 years since it was built by the Romans in 122 C.E. This convinced Hutton that the processes that wore down entire mountains must be incredibly slow. (Photo by the author)

Investigation of the Laws Observable in the Composition, Dissolution, and Restoration of Land Upon the Globe." Finally, in 1795 he published a book-length version of *Theory of the Earth*.

Hutton's ideas were astonishing and remarkably advanced for their time. By the late 1700s, scholars knew a lot about rocks, strata, and fossils, but there was no general theory of geology. One of the impediments was the still widely held belief that the earth had been created only about 6,000 years earlier, according to the Ussher-Lightfoot interpretation of the book of Genesis. Some geologists thought that sedimentary rocks were formed when immense quantities of minerals precipitated out of the waters from Noah's flood. Many scholars realized that the destructive processes of erosion were important, but there was no equivalent explanation for the uplift and creation of the landscape.

Hutton gained his greatest insights on how long this process must have taken when he found outcrops of what is now known as an angular unconformity (figure 4.3). To Hutton, an angular unconformity was proof of the immense age of the earth. He saw the tilted layers at the bottom as having once been deposited horizontally on the bed of a stream or the floor of the ocean, then hardened into sandstones and shales, and then tilted vertically by immensely strong forces. The sharp erosional surface cutting through the lower tilted beds must represent their uplift into mountain ranges, and subsequent erosion wearing them down again over millions of years. Finally, the horizontal beds on top represent the accumulation of yet another long piling up of sediments on the bottom of a river or the ocean, which takes millions of years if you assume modern rates of sedimentation. All together, any singular angular unconformity must represent millions of years of time at the minimum, not the 6,000 years that the Bible supposedly indicated.

In 1787, Hutton discovered an angular unconformity in the east bank of the Jed Water, just south of the town of Jedburgh (figure 4.3A). As he wrote in 1795,

> I was surprised with the appearance of vertical strata in the bed of the river, where I was certain that the banks were composed of horizontal strata. I was soon satisfied with regard to this phenomenon, and rejoiced at my good fortune in stumbling upon an object so interesting to the natural history of the earth, and which I had been long looking for in vain. . . . above those vertical strata, are placed the horizontal beds, which extend along the whole country.

A

B

Hutton continued to explore outcrops around Scotland (see chapter 5) for evidence to support his views later published in *Theory of the Earth*. He found additional examples of angular unconformities at Teviotdale and on the Isle of Arran, but the exposures were poor and not easy to date. But on his last field excursion in 1788, he took a small boat along the Berwickshire coast with his friends and followers James Hall and John Playfair. He knew that coming down the coast southeast from Edinburgh, the outcrops were the vertically tilted sandstones and shales then called the "schistus" (now known to be Silurian in age, 435 million years old), also exposed below the unconformity in Jedburgh. But if you came up from the south, the outcrops were mainly horizontal beds of the Old Red Sandstone (now known to be Late Devonian, about 370 million years in age). Like any good detective (or geologist), Hutton knew there must be some place along the coast where the two must meet. Finally, he found it at Siccar Point (figure 4.3B).

As Playfair wrote about that momentous day:

> On us who saw these phenomenon [*sic*] for the first time the impression will not easily be forgotten . . . We felt necessarily carried back to a time when the schistus on which we stood was yet at the bottom of the sea, and when the sandstone before us was only beginning to be deposited, in the shape of sand or mud, from the waters of the supercontinent ocean . . . The mind seemed to grow giddy by looking so far back into the abyss of time; and whilst we listened with earnestness and admiration to the philosopher who was now unfolding to us the order and series of these wonderful events, we became sensible how much further reason may sometimes go than imagination may venture to follow.

Hutton's ideas were a huge departure from contemporary geological thinking. He asserted that sedimentary rocks were once sands and muds that had been washed off the land from rivers and into the oceans, accumulated in beds there, and then solidified into rocks. But he argued that the

Figure 4.3 ◄

Angular unconformities in Scotland: (*A*) At Inchbonny in the valley of the Jed Water, south of Jedburgh, showing the way it was drawn by John Clerk for Hutton's book. The steeply tilted Silurian "schistus" lies beneath an erosional surface that cut across the sandstones and shales after they were tilted. They were then covered by the nearly horizontal beds of the Devonian Old Red Sandstone. (*B*) The famous angular unconformity at Siccar Point. The beds show the same Silurian "schistus" overlain by Devonian Old Red Sandstone as in Jedburgh. (Courtesy of Wikimedia Commons) hardening into rocks was due not to the simple precipitation of sand and mud out of a watery solution but rather was due to the effects of pressure and heat, which is what modern geology has confirmed.

Hutton claimed that the totality of these geologic processes could fully explain the current landforms all over the world, and no biblical explanations were necessary in this regard. Finally, he argued that the processes of erosion, deposition, sedimentation, and uplift of mountain ranges were cyclical and must have been repeated many times in Earth's history. Given the enormous spans of time taken by such cycles, Hutton asserted that the age of the earth must be inconceivably great. As Hutton himself wrote, geologic time is immense and virtually endless, with "no vestige of a beginning, no prospect of an end." As Stephen Jay Gould wrote, Hutton "burst the boundaries of time, thereby establishing geology's most distinctive and transforming contribution to human thought—Deep Time."

FOR FURTHER READING

Broadie, Alexander, ed. *The Scottish Enlightenment: An Anthology*. London: Canongate Classics, 2008.

Buchan, James. *Crowded with Genius: Edinburgh, 1745–1789*. New York: Harper Collins, 2009.

Geikie, Archibald. *James Hutton: Scottish Geologist*. Shamrock Eden Digital Publishing, 2011.

Herman, Arthur. *How the Scots Invented the Modern World: The True Story of How Western Europe's Poorest Nation Created Our World and Everything in It*. New York: Broadway Books, 2007.

Hutton, James. *Theory of the Earth with Proofs and Illustrations*. Amazon Digital Services, 1788.

McIntyre, Donald B., and Alan McKirdy. *James Hutton: The Founder of Modern Geology*. Edinburgh: National Museum of Scotland Press, 2012.

Repcheck, Jack. *The Man Who Found Time: James Hutton and the Discovery of the Earth's Antiquity*. New York: Basic Books, 2008.

Rudwick, Martin J. S. *Earth's Deep History: How It Was Discovered and Why It Matters*. Chicago: University of Chicago Press, 2014.

05 | THE "EARTH'S GREAT HEAT ENGINE": THE ORIGIN OF MAGMAS

IGNEOUS DIKES

The volcano was not created to scare superstitious minds and plunge them into fits of piety and devotion. It should be seen as the vent of a furnace.

—JAMES HUTTON

NEPTUNISM VERSUS PLUTONISM

In the late 1700s, the early natural historians of the Enlightenment were still strongly influenced by the Genesis accounts and the myth of Noah's flood. Early scholars such as Giovanni Arduino (1714–1795) tried to shoehorn all the rocks they could see into a simplistic sequence of hard crystalline "Primary" or "Primitive" rocks (granitic rocks and metamorphic rocks like schist or gneiss), supposedly formed when the earth was created. These were covered by "Secondary" rocks, which were fossiliferous layered sedimentary beds, often folded and deformed. (Today, most of these "Secondary" rock strata are assigned to the Devonian through Cretaceous Periods.) According to some naturalists, the "Secondary" beds were the main deposits formed by Noah's flood. Above these were loosely consolidated sediments and sedimentary rocks called "Tertiary" beds, which were supposedly post-flood deposits.

These simplistic ideas of the rock record could be entertained as long as you didn't go out in the real world and look at real rocks too closely. But most geologists at that time did not travel very far or check their ideas against the outcrops, instead forcing the limited rock exposures of northern Europe into their previously accepted dogmas. This scheme of all the rocks supposedly formed by water (usually interpreted as Noah's flood) became known as Neptunism, after the Roman god of the sea. The Neptunists argued that even lava flows were once laid down in water. Their opponents, however, argued that lava flows were once formed of hot molten rock, not from water, and so were called "Plutonists" (after Pluto, the Roman god of the hot volcanic underworld).

Among the most prominent Neptunists was German naturalist Abraham Gottlob Werner (1749–1817). A professor of mineralogy at the Freiburg Mining Academy, Werner was said to be a spellbinding lecturer and a powerful personality who made devoted converts of nearly anyone who heard him speak. His concepts were among the most popular in Europe, primarily because of the force of his arguments and personality, and were not based on close examination of outcrops over a wide area. (He did not, however, specify that layered sedimentary rocks and lava flows were deposits of Noah's flood, as many of his contemporaries believed, but simply that they were laid down in water.) His disciples were found in all the major European universities, including the University of Edinburgh, where Robert Jameson was a convinced Neptunist who also became the archrival of James Hutton. Even the great poet and naturalist Goethe was a convinced Neptunist. In the fourth act of *Faust*, there is a dialogue between a Neptunist and a Plutonist, wherein Mephistopheles is clearly a spokesperson for the evil Plutonist viewpoint.

You might ask yourself, "How could anyone think a volcanic rock was formed in water?" Remember, chemistry was in its infancy at this time, so no one knew anything about the heat and pressure needed to melt rocks. Few Europeans had ever seen a lava flow. We are used to watching videos of glowing hot molten lava flowing in active volcanoes like Kilauea, but in Hutton's day, few Europeans traveled far from their hometowns, and none had seen a volcano erupt unless they had been to southern Italy at a time when Vesuvius, Stromboli, or Etna might be erupting—and those volcanoes mainly produce ash, not lava flows. It was a French geologist, Nicolas Desmarest (1725–1815), who pointed out in 1774 out that the extinct volcanoes of Auvergne in southern France showed all the evidence of having once been actively erupting volcanoes, based on their volcanic cones and many weathered lava flows. This evidence alone should have proved the case for Plutonism, although the Neptunism were the dominant theory for many more years.

Yet as James Hutton thought about his grand scheme of mountains being uplifted and eroded away, he became convinced that rocks such as granites and the basalt that formed lava flows were produced by hot molten rocks known as magma, not laid down in water. But there were no active volcanoes anywhere in northern Europe, and Hutton had never seen a lava flow move across the landscape. In the absence of this evidence, Hutton looked for places where granitic rocks or basalts had melted their way, or intruded, through preexisting rock and baked the rock surrounding them with their enormous heat.

GLEN TILT AND SALISBURY CRAGS

Hutton got his first clues when he noticed that the gravels in the River Tilt, flowing south out of the Cairngorm Mountains in the Highlands north of Edinburgh, were full of both granitic pebbles and older metamorphic rocks. From this, he knew that both kinds of rocks must be exposed in the riverbed, and he might possibly see where one contacted the other. In 1785 he rode up the valley of the Tilt and stayed overnight at Forest Lodge. The next day, exploring the exposed rocks in the bed of the River Tilt at Dail-an-eas Bridge near Forest Lodge, he found what he had been looking for: veins of brick-red granite cutting through older metamorphic rocks and baking the zones around them (figure 5.1). Here was proof that granites were once molten magmas, not formed in water! Not only that, but the granites must be younger than the schists, and therefore not all had been formed during the original Creation of the earth as described in Genesis.

But Hutton needed something even more convincing: lava flows that were intruded through layered sediments that had been formed in water. As Hutton wandered around the hills south of Edinburgh with his dog Missy, he also came to realize that the mountain that towers above the city, Arthur's Seat, was the throat of an old volcano (figure 5.2), and the Salisbury Crags on the north flank of the mountain are a ledge of old volcanic rock. Finally, on the southwestern slopes of the Crags, he found what he was looking for: intrusions of basaltic lava that had melted their way through layered sediments and even deformed them in the process (figure 5.3). This spot is so famous that it is now known as "Hutton's Section," and geology classes visit it on a regular basis. In 1786 Hutton found another example in Galloway, and in 1787 he located yet another on the Isle of Arran.

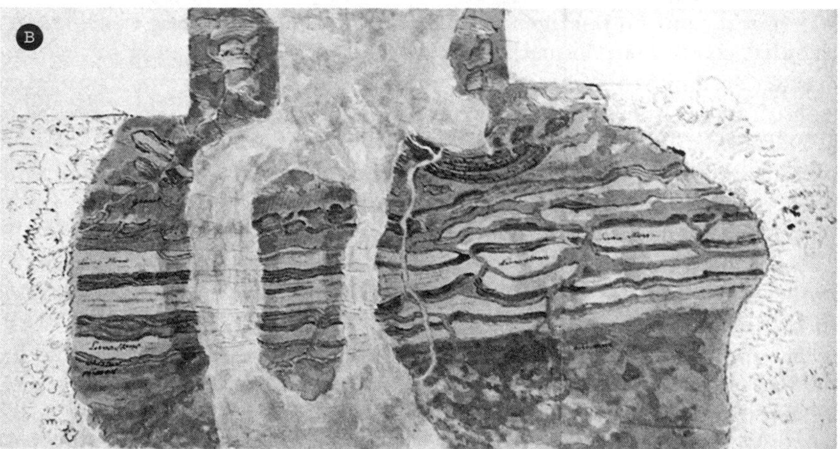

Figure 5.1 ▲

The granitic intrusions at Glen Tilt that helped confirm James Hutton's Plutonist viewpoint of the earth: (*A*) The outcrop as it is today above Dail-an-eas Bridge, looking northeast up Glen Tilt, showing the white veins of granites intruding through the dark Precambrian schists of the Scottish Highlands. (*B*) An illustration drawn by John Clerk from a posthumous edition of Hutton's book, showing the dikes intruding through the older rocks. ([*A*] Courtesy of British Geological Survey; [*B*] Courtesy of Wikimedia Commons)

Figure 5.2 ▲
Arthur's Seat, the throat of an extinct Carboniferous volcano, towers above the south side of Edinburgh. The ledges in the foreground are the Salisbury Crags, volcanic sills that intruded parallel to the bedding in the Carboniferous sedimentary country rocks. Hutton lived in a house at the base of the Salibury Crags to the right in this shot, and frequently hiked in the area with his dog Missy. This shot is taken from near Edinburgh Castle, which sits on another vent of the same volcano. (Courtesy of Wikimedia Commons)

Hutton and his chemist friends, Joseph Black and Sir James Hall among them, also had much better insights into the chemistry of rocks than anyone else at that time. Hutton knew what minerals (such as salt) formed by chemical precipitation from water looked like, and he could see that magmas were not formed in water. When Hutton moved to Edinburgh in 1768, he worked with Black, who shared his love of chemistry, a key tool to understanding the effect of heat on rock. Black deduced the existence of latent heat and the importance of pressure on heated substances. Water, for instance, stays liquid under pressure, even when heated to a temperature that normally would transform it to steam. Those ideas about heat and pressure would become key to Hutton's theory about how buried sediments became rock. In 1792 Hall performed an experiment in which he melted

Figure 5.3 ▲

Hutton's Section at the Salisbury Crags: (*A*) The outcrop as it looks today, showing the baked and deformed sedimentary layers (bottom) with magma surrounding them (top). (*B*) Close-up of the right side of the outcrop, showing the bent sedimentary layers surrounded by intruding magma. (*C*) A diagram drawn by John Clerk and published in a posthumous edition of Hutton's book with the same outcrop rendered on a completely wrong scale. ([*A*] Photo by the author; [*B*] Photo by the author; [*C*] Courtesy of Wikimedia Commons) a piece of basalt at temperatures between 800°C and 1,200°C, and it recrystallized into basalt again when it was cooled slowly. This was one of the first examples of experimentation in geology and proved what molten rocks would look like in nature.

THE DYNAMIC EARTH

Having read accounts of hot springs and volcanoes (but never having visited them), Hutton was convinced that the earth was hot and molten in its center, powered by what he called "the Earth's great heat engine." In his words, "The volcano was not created to scare superstitious minds and plunge them into fits of piety and devotion. It should be seen as the vent of a furnace." His ideas were further confirmed by coal seams that had been baked by lava that had intruded through them. He believed that this heat engine was responsible for the uplift and building of mountains, which then became the sediments that were washed to the sea. These processes went on in an endless cycle of uplift, erosion, deposition, and uplift again. All of these ideas were part of a broad overview of a dynamic Earth that was incredibly old and constantly remade and recycled—not a young Earth that had remained essentially unchanged since its creation 6,000 years earlier.

With the publication of his essays in 1788 and his book *Theory of the Earth* in 1795, Hutton put forth his concepts for the rest of the world to see. Yet Hutton's ideas were not immediately accepted, partly because his writing was so difficult to read and understand. When Hutton died in 1797, he was still not widely appreciated, although in 1802 his protégé, John Playfair, published *Illustrations of the Huttonian Theory of the Earth*, which made the case clearer and was more widely read (and had illustrations by Hutton's friend John Clerk, which further helped people understand what Hutton had seen).

It would take another generation for such revolutionary ideas to win acceptance in the geological community. This happened thanks to a young man named Charles Lyell (figure 5.4), born in 1797, the year Hutton died. Originally trained in the law to become a barrister, he soon became bored and instead pursued the young field of geology as a hobby. He traveled widely over Europe, witnessing many different geological phenomena with the uniformitarian eyes of Hutton. Eventually, Lyell wrote his masterpiece, *Principles of Geology*, published in three volumes from 1830 to 1833.

Figure 5.4 ▲
Charles Lyell, in his later years, after he had been knighted and had become one of the most respected figures in all of science due to his books on uniformitarian approaches to geology. (Courtesy of Wikimedia Commons)

The book is written essentially as a legal brief (which as any lawyer can tell you, is never actually "brief"). Mustering all the observations he had gathered from his travels and his reading, he used his skills as a lawyer to argue a decisive case for the uniformitarian view of the earth. Like any good lawyer, he used any tactics necessary to discredit his opponents, the catastrophists, while presenting overwhelming evidence for his own case. He confirmed Hutton's ideas of "Earth's great heat engine" with multiple accounts of volcanic eruptions and hot springs, especially in southern Italy. Within a few years, the last of the old-line catastrophists and Neptunists had died or given up, and geology became a modern science.

FOR FURTHER READING

Bonney, Thomas G. *Charles Lyell and Modern Geology*. New York: Andesite, 2015.

Geikie, Archibald. *James Hutton: Scottish Geologist*. Shamrock Eden Digital Publishing, 2011.

Hutton, James. *Theory of the Earth with Proofs and Illustrations*. Amazon Digital Services, 1788.

Lyell, Charles. *Principles of Geology*. 3 vols. Chicago: University of Chicago Press, 1990–1991.

McIntyre, Donald B., and Alan McKirdy. *James Hutton: The Founder of Modern Geology*. Edinburgh: National Museum of Scotland Press, 2012.

Repcheck, Jack. *The Man Who Found Time: James Hutton and the Discovery of the Earth's Antiquity*. New York: Basic Books, 2008.

Rudwick, Martin J. S. *Earth's Deep History: How It Was Discovered and Why It Matters*. Chicago: University of Chicago Press, 2014.

06 | THE ROCK THAT BURNS FIRES THE INDUSTRIAL REVOLUTION

COAL

My name's Polly Parker, I come o'er from Worsley,
My mother and father work down the coal mine.
Our family is large, we have got seven children,
So I am obliged to work down the same mine.
And as this is my fortune I know you'll feel sorry
That in such employment my days I must pass.
But I keep up my spirits, I sing and look cheerful,
Although I am but a poor collier lass.

By the greatest of dangers each day I'm surrounded
I hang in the air by a rope or a chain.
The mine may give in, I may be killed or wounded
Or perish by damp or the fire of a flame.
But what would you do if it weren't for our labours
In greatest starvation your days you would pass,
For we would provide you with life's greatest blessing
So do not despise a poor collier lass.

All the day long you may say we are buried,
Deprived of the light and the warmth of the sun.
And often at night from our beds we are hurried,
The water is in and barefoot we run.
And though we go ragged and black are our faces,
As kind and as free as the best we'll be found.
And our hearts are more wide than your lords' in high places
Although we're poor colliers that work underground.

—"COLLIER'S LASS," TRADITIONAL MINER'S SONG

A LUMP OF COAL

People have been digging coal from the ground since as early as 4000 B.C.E. in China, where it was mostly used to create heat in fireplaces and furnaces.

But by about 1000 B.C.E., the Chinese were using coal to smelt copper. When Marco Polo made his historic visit to China from 1271 to 1295, he came back with tales of how the Chinese used "black stones . . . which burn like logs." He was amazed to see that coal was so plentiful that people could take three hot baths a week.

However, coal had been used in Europe since ancient times, and its uses were merely forgotten in the Middle Ages when Marco Polo lived. Around 300 B.C.E., the Greek philosopher Theophrastus described it in his geological treatise *On Stones*:

> Among the materials that are dug because they are useful, those known as *anthrakes* [coals] are made of earth, and, once set on fire, they burn like charcoal. They are found in Liguria . . . and in Elis as one approaches Olympia by the mountain road; and they are used by those who work in metals.

Evidence of coal in funeral pyres has been found in Bronze Age English sites dated to 3000 B.C.E. The Romans had excavations in most of the coalfields of England, Scotland, and Wales by 200 C.E. Coal was used not only in heating furnaces and smelting, but also for hearths and for central heating in villas, and for heating the water in the Roman baths, such as the famous one in Bath, England.

THE INDUSTRIAL REVOLUTION

Through the Middle Ages and up until about 1700, coal was still a minor resource, since it was difficult to mine and there was abundant wood for charcoal and other types of fuel. All of this changed when the Industrial Revolution began to accelerate in the late 1700s. Although waterwheels and other sources of power were widely used, there were not enough rivers to power large factories, and they would have run out of suitable river sites in England by 1830. Thus, the invention of the workable steam engine in the late 1700s supplied the most practical source for a large amount of power to run a factory or propel a ship or locomotive. Small steam engines could be heated with wood-burning furnaces, but a cheaper, more concentrated source of energy was needed to power larger machines, and coal took over as the first great fuel of the Industrial Revolution and made the entire Industrial Age possible (figure 6.1).

Since the Industrial Revolution grew up in Great Britain, that country was among the first to undertake large-scale coal mining. By 1800, 83 percent of

Figure 6.1 ▲
An old lithograph showing the hard work of coal mining. (Courtesy of Wikimedia Commons)

the world's coal was mined in Britain, especially in the huge coalfields in southern Wales and in central and northern England from Manchester to Newcastle and into southern Scotland (figure 6.2). At its peak in 1947, the British coal industry had about 750,000 miners in dozens of mines around the country (figure 6.3).

However, mining for coal was a dirty, dangerous, and often deadly business. In the early days of coal mining, the owners had all the power, and the workers had to accept whatever mine conditions prevailed—or else they would starve. And those conditions were horrific. Coal mining often releases lots of fumes that can be either toxic or explosive or both, so coal-mining explosions were a common problem for years. Coal miners would take a caged bird (typically a canary) into the coal mine with them, because the bird was more sensitive to gases and would react before the miners could detect the fumes (hence the phrase "canary in a coal mine" for something that warns us of upcoming problems). Coal mining also produced huge amounts of black coal dust, which filled the miners' lungs, so

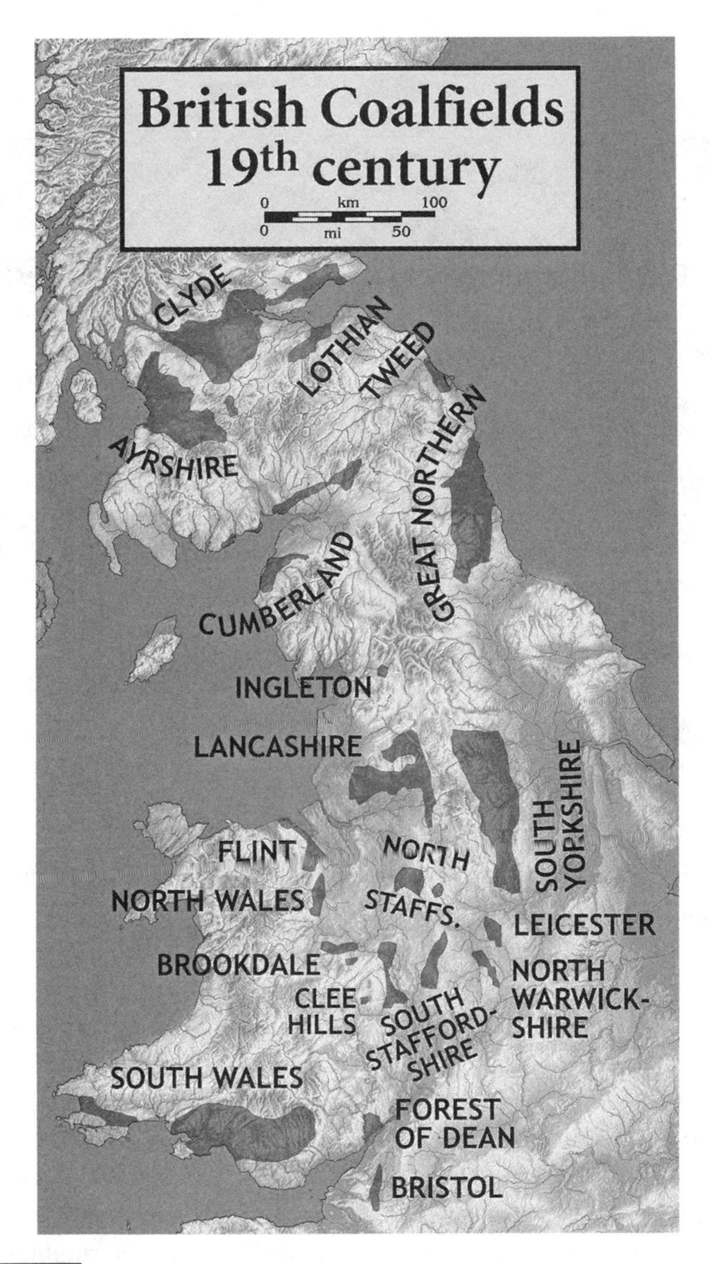

Figure 6.2 ▲

The location of the major coalfields in Great Britain in the nineteenth century. (Courtesy of Wikimedia Commons)

Figure 6.3 ▲
Archival photograph of Irish coal miners and their ponies, circa 1884. (Courtesy of Wikimedia Commons)

many died of black lung disease. Coal mines were also prone to caving in, often burying hundreds of miners alive.

Even more appalling was the fact that children as young as 8 years of age were expected to work in the coal mines in the nineteenth century (figure 6.4). With their smaller size, they could work in more confined spaces. They were particularly important for opening and closing trapdoors in the mines to let the mine carts through while preventing the gases from building up. In the eighteenth and nineteenth century, they would work 12-hour shifts in the mine, just like the men, 6 days a week, with only Sundays off. They would sit in pitch darkness most of that time, only lighting a candle when necessary, and listen for the rumble of the mine carts to warn them to open the doors before the carts came through. During the cold, short days of winter, these children would get up in the dark, work in the dark for 12 hours, and come home after dark, so they only saw daylight on Sundays.

Figure 6.4 ▲
Archival photograph of children working in West Virginia coal mines, driving the ponies, circa 1908. (Courtesy of Wikimedia Commons)

The casualties of mining were appalling. In the United States alone, more than 90,000 miners died between 1900 and 1950, with 3,200 dying in 1907 alone. Even with modern safety regulations, about 28 miners per year died between 2005 and 2014, making it one of the most hazardous jobs. If a miner did not die suddenly from an explosion, cave-in, or fire, he still died young from black lung disease. Thanks to the hard work of the labor unions, miners gradually began to win concessions from the coal barons in the twentieth century, and eventually laws were passed that mandated safety, reduced the hours of each shift, and outlawed child labor.

As the Industrial Revolution spread to other parts of the world, huge coal deposits were discovered to support the rapid rush to industrialization. In the United States, major coal deposits were found in the Appalachian region of western Pennsylvania, Virginia, West Virginia, and adjacent areas of Kentucky, Ohio, and Tennessee. By 1870 these fields were producing 40 million tons of coal, and then production doubled every 10 years.

By 1900 the number had jumped to 270 million tons, and then it peaked at 680 million tons in 1918 as huge demands for coal to power ships and factories during World War I pushed production.

Germany had a similar area of coal deposits in the Ruhr Valley, where nearby iron deposits combined to make it an industrial powerhouse. In 1850, the average mine only employed 64 men and produced about 8,500 short tonnes, for a total output of 2 million tons. By 1900, these same mines each produced 280,000 short tonnes and employed 1,400 men, for a total output of 60 million short tonnes. Coal deposits were also found in many other European countries, including France, Belgium, Austria, Hungary, Spain, Poland, and Russia. Eventually coal mining spread around the world, so major deposits in Russia, India, Japan, Australia, New Zealand, and South Africa were intensively developed by 1900. Today, China is the leading coal-producing country in the world, with over 2.8 billion tons produced in 2008, about 40 percent of all the world's coal production. However, many other countries have depleted their deposits, or their coal is too high in sulfur to mine without causing acid rain, or the deposits are too expensive to mine in the face of competition from cheaper sources of energy.

THE "COAL MEASURES"

The search for coal was not only of great economic importance to the Industrial Revolution, but it was also the foundation for some of the earliest geological studies in Britain or anywhere in the world. As soon as people began to study the major coalfields, they recognized that there was a certain sequence of rocks that held most of Britain's coal. Called the "Coal Measures" in Britain by the early 1700s, it became the basis for the geologic time term "Carboniferous" (meaning "coal bearing"), formally named by William Conybeare and William Phillips in 1822, almost a century later.

One of these pioneers of geology was John Strachey (1671–1743), a squire from Somerset who was interested in the coal works near and under his estate. In 1719 he published a famous diagram (figure 6.5) that shows one of the first true geological cross sections ever drawn. He first mapped the coal seams on the surface and measured their thickness and angle of dip into the ground. Then he correctly projected their plunge into the subsurface and showed what they must be doing in three dimensions. From this diagram, he could not only establish his rights to a coal lease, but he could

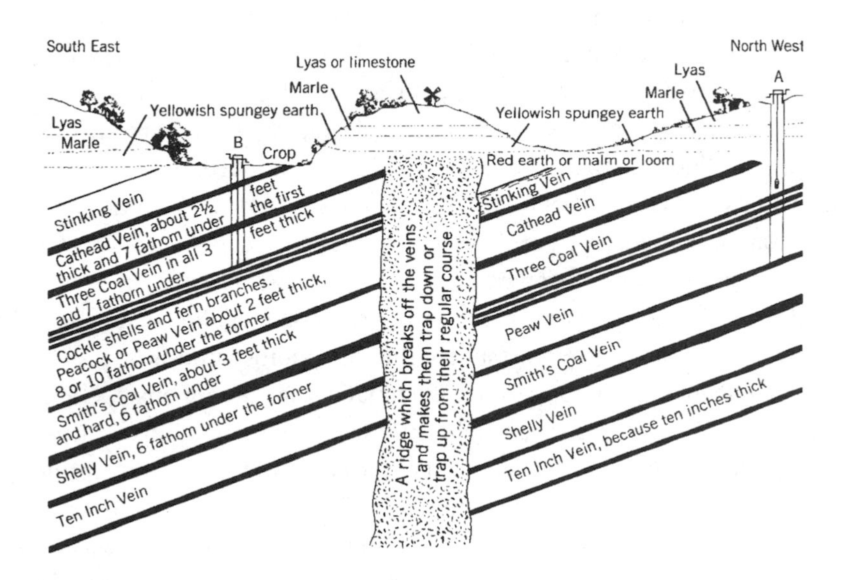

Figure 6.5 ▲
John Strachey's famous pioneering geological cross section of the coalfields in his area. (Courtesy of Wikimedia Commons)

predict the location and thickness and potential amount of coal in a given area, as well as where it should be found. This was a gigantic leap from simply locating a seam of coal exposed on the surface and following the seam downward by digging, as generations of miners had long done. As we shall discuss in chapter 7, it was crucial to William Smith's insights around 70 years later about the stratigraphic sequence in not only Somerset but all of Great Britain.

But why were there such extensive coal deposits around the world in the Carboniferous and not as many in any other time period? A number of geologic events interacted in a unique way. First, there were no tree-sized land plants before the Late Devonian, but in the Carboniferous they began to evolve into huge club moss trees, gigantic horsetails, and dense forests of ferns. These plants grew in areas of dense swampy vegetation that formed on the floodplains, river deltas, and coastal lagoons along the fringes of the newly formed Appalachian Mountain Range in North America, and similar mountains in Eurasia created by the collision of the various continents to form Pangea.

As these giant swamps grew in the tropical regions of Eurasia and North America, they were unlike any swamps that have grown since. Unlike modern swamps, where there are many termites and other decomposers to quickly break down the trees as they die and sink in the stagnant water, no such wood-digesting insects had yet evolved in the Carboniferous. Thus, enormous volumes of vegetation sank into the stagnant acidic muds of the coal swamps and became permanently buried in the earth's crust, instead of rotting away, as they do today.

So much coal accumulated in the earth's crust that it pulled many tons of carbon dioxide out of the atmosphere through photosynthesis and locked up a significant volume of carbon. Eventually, the "greenhouse" climate of the Early Carboniferous—no polar ice caps, high carbon dioxide levels, and high sea levels that drowned most of the continents—was transformed into an "icehouse" planet by the Late Carboniferous—polar ice caps on the South Pole, lower carbon dioxide levels, and much lower sea levels as all that polar ice pulled water out of the ocean basins. Earth was dominated by this "icehouse" planet setting for almost 150 million years.

The earth has switched back and forth between "greenhouse" and "icehouse" stages many times over the past billion years. Unlike the runaway greenhouse that is Venus (with an atmosphere full of sulfuric acid, hot enough to melt lead) or the frozen iceball that is Mars, Earth has life that regulates its carbon cycle. Carbon is locked up in the crust in the form of limestones (made mostly by seashells and other fossils) and coal (produced by plants). Earth's living systems act as a thermostat, preventing the planet from a runaway greenhouse or a runaway icehouse.

THE CURSE OF COAL

Unfortunately, the very coal mining that powered the Industrial Revolution is now destroying the habitability of our planet. Since the 1700s, we have mined and released many million tons of carbon dioxide once locked into the earth's crust by burning millions of tons of coal. All this carbon dioxide, once trapped in the crust by the unique conditions of the Carboniferous Period, is now creating a planetary "super-greenhouse" faster than ever seen in the geological past. Those pioneers who first mined coal to power their steam engines unwittingly upset the delicate balance of carbon in the earth's atmosphere, oceans, and crust.

In addition to being one of the worst greenhouse gas producers among all fossil fuels, coal has many other environmental hazards as well. Shaft mining to reach deep coal seams was always dangerous to the miners and leaves huge piles of tailing and mine waste and pools of toxic sludge, as well as shattered lives. Even more destructive is strip mining, where huge amounts of soil and rock "overburden" are removed to reach a coal layer below the surface. This destroys vast areas of landscape at a major cost to the environment. In the early days, strip miners just left huge spoil piles on the ground, separated by the flooded mining pits, turning the landscape into a scarred wasteland. Since the advent of environmental regulations, mining companies now must put the waste rock back after the coal has been removed, and then restore it to farmland or whatever it was before it was disturbed. In many cases, it makes the strip mining pointless, because the coal doesn't earn enough money to compensate for the huge cost of mining and restoration. Lately, the practice has been "hilltop removal," where the mining company takes the entire top off a hill to reach a coal seam, then dumps the waste rock into a valley and completely transforms the landscape.

Another environmental cost of coal mining is acid rain. Coal that has lots of sulfur in it burns to form sulfuric acid, which then is blown downwind of the power plant and kills everything that it rains upon. Acid rain nearly killed the Black Forest of southern Germany, and badly damaged the forests of the northeastern United States. Again, environmental regulations finally came along with the Clean Air Act of 1970 that put a high cost on mining high-sulfur coal, so many mines in Appalachia and Illinois closed down because they were not cost-effective. Instead, the low-sulfur coals of the Powder River Basin in Wyoming were much cheaper to mine. Now there is a cap-and-trade system in place to discourage burning high-sulfur coal, and its environmental damage has declined dramatically.

For all these reasons, coal is considered one of the most environmentally destructive forms of fossil fuel and the most hazardous to miners and other people. Regulations have reduced coal's acid rain impact, limited its destruction of the landscape, and made the miner's job less deadly, but coal is still a huge producer of greenhouse gases. Many environmentalists have fought for a long time to find ways to gradually phase it out completely and replace it with other cleaner sources of energy.

Ironically, it was not regulations, but Adam Smith's "invisible hand" of the capitalist free market that largely accomplished this task. The breakthroughs in solar technology and cheap solar electricity and wind power, the plunge in the price of oil during the years from 2014 to 2017, and especially the oversupply of natural gas, drove down energy prices until coal was no longer competitive in most parts of the world. In 2016, Peabody Energy, the largest coal company in North America, filed for bankruptcy, and coal mining has almost ceased in the eastern United States. The same forces have nearly completely shut down coal production in the United Kingdom, so that all that remains of the once vast British and Welsh coalfields are closed mines and a devastated landscape. The sole remaining country to mine and burn a lot of coal is China, now the largest coal producer in the world. However, China is also trying to phase out coal mining and coal-fired power plants because of the poisonous air pollution they produce, and it has already made major steps in that direction.

Coal was the fuel that powered the Industrial Revolution and built the modern world. It was also a resource that trapped much of the earth's carbon in the crust, and with the burning of this coal, we unleashed another greenhouse world on future generations. Fortunately, the Age of Coal appears to be coming to an end, but it is an open question whether we can cut back enough on our burning of all fossil fuels to avoid disaster.

FOR FURTHER READING

Berry, William B. N. *The Growth of a Prehistoric Time Scale*. San Francisco: Freeman, 1968.

Freese, Barvara. *Coal: A Human History*. New York: Penguin, 2004.

Goodell, Jeff. *Big Coal: The Dirty Secret Behind America's Energy Future*. New York: Mariner, 2007.

Martin, Richard. *Coal Wars: The Future of Energy and the Fate of the Planet*. New York: St. Martin's, 2015.

Thomas, Larry. *Coal Geology*, 2nd ed. New York: Wiley-Blackwell, 2012.

07 THE MAP THAT CHANGED THE WORLD: WILLIAM SMITH AND THE ROCKS OF BRITAIN

JURASSIC WORLD

Organized Fossils are to the naturalist as coins to the antiquary; they are the antiquities of the earth; and very distinctly show its gradual regular formation, with the various changes inhabitants in the watery element.

—WILLIAM SMITH

A SLICE THROUGH THE EARTH

Abraham Gottlob Werner, James Hutton, and most naturalists of the late 1700s were focused on a broadscale theoretical understanding of the earth. They worked with limited outcrops in Scotland, Germany, and elsewhere, and generalized about the large-scale history of the earth based on limited evidence. Some of them got it wrong (Werner), while others were generally right (Hutton). They were all wealthy gentlemen who didn't have to work for a living, or learned professors with secure positions, and could spend their time as they wished. They had the education, money, and free time needed to study geology, which was just a hobby and not a profession.

However, the explosive growth of coal mining in the late 1700s created a new need to understand the earth in a more practical, detailed, local manner. As we saw in chapter 6, people searching for coal needed to map the distribution of coal seams and predict where more coal could be found. The pioneering geological cross section of the Somerset coal deposits made by John Strachey in 1719 (see figure 6.5) was the first attempt to draw a cross section through the earth's shallow crust and visualize it at depth. Still, most naturalists of the time were not interested in such painstaking work as mapping and drawing cross sections, preferring to generalize about large-scale theoretical models of the earth from their armchairs rather than study the outcrops in detail. In addition, the landscape of Great Britain and most of Europe is heavily vegetated, so there are very few outcrops, and it's difficult to get a sense of the geology below your feet in most places.

William Smith (1769–1839) was not a rich British gentleman like most of the founders of geology (figure 7.1). The son of a blacksmith—John Smith, who died when young William was just 8 years old—Smith had none of the advantages of the upper class. Nonetheless, he was very bright and industrious. He taught himself far beyond his limited schooling, showing

Figure 7.1 ▲
Portrait of William Smith. (Courtesy of Wikimedia Commons)

a special aptitude for mathematics and drawing. At the age of 18, he was apprenticed to Edward Webb, a surveyor working in Gloucestershire. He quickly became an excellent surveyor, able to work on almost any project that came along.

In 1791 he was hired to work on Sutton Court in Somerset, the same estate where Strachey had surveyed and prospected for coal more than 70 years earlier. There he learned about Strachey's maps and cross sections, which influenced his own thinking as he surveyed routes across the countryside. He and Webb worked on that project for 8 years, finding routes for canal excavations in Somerset, especially the Somerset Coal Canal (figure 7.2). At that time, there were a lot of canals being dug across England, since the Industrial Revolution required a cheap form of transportation for all the coal and other commodities that had to be carried to the great industrial cities. Smith had the unusual opportunity of seeing the fresh bedrock exposures through much of England in areas normally covered by vegetation and difficult to map. He also surveyed and studied many of the coal mines in the region and saw more cross sections through the earth's crust than any Briton before him.

Most of the rocks of Oxfordshire and Somersetshire are part of the famous Jurassic sequence (figure 7.3) that yielded so many fossils of marine reptiles, ammonites, and other sea life down on the coast at Lyme Regis. These rock units are now legendary not only for their fossils, but also for their colorful and quaint names: "Blue Lias" (the main fossiliferous layer at Lyme Regis), "Shales with Beef," "Black Ven Marls," "Green Ammonite Beds," "Cornbrash," "Corallian Group," "Inferior Oolite," "Forest Marble," "Kimmeridge," and the famous "Oxford Clay," which yielded not only invertebrate fossils, but also many more marine reptiles. (Some of these names, such as Cornbrash, were coined by William Smith himself and are still used today.) The Jurassic Taynton Limestone northeast of Oxford produced fossils of the first dinosaur ever named, *Megalosaurus*, although this fossil was not formally named until 1824, and the concept of "dinosaurs" did not emerge until the 1830s and 1840s.

FAUNAL SUCCESSION

As he saw the Jurassic stratigraphic sequence of western England repeated over and over again, he understood not only the formations but also the

Figure 7.2 ▲

(*A*) Modern remains of the Somerset Coal Canal, first surveyed by William Smith in the 1790s. (*B*) The William Smith House in Tucking Mill, just south of Bath, where he lived during his work on the Somerset Coal Canal. It is the only building remaining in which Smith lived. (Photos by the author)

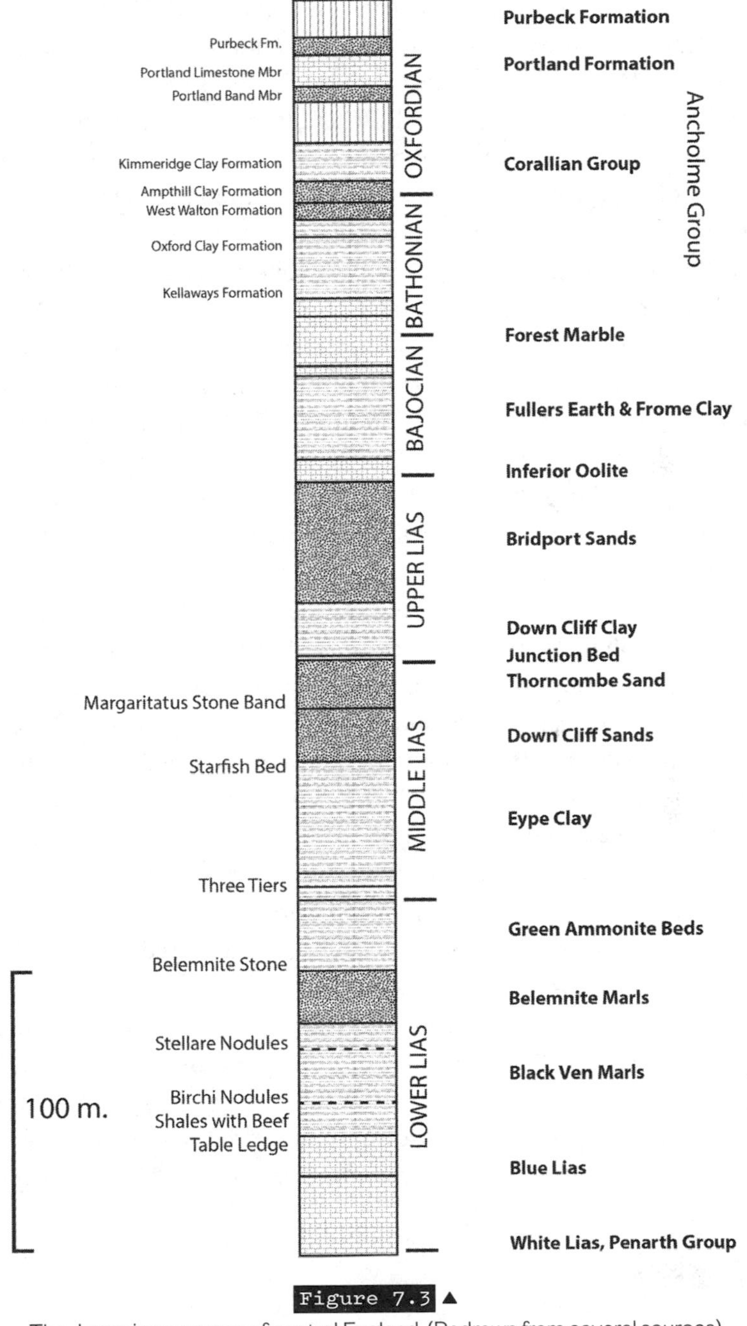

Figure 7.3 ▲

The Jurassic sequence of central England. (Redrawn from several sources) characteristic fossils found within each unit (figure 7.4). More importantly, he realized that similar-looking rock units could best be distinguished by their fossils. This is known as the principle of faunal succession, and it is the foundation for biostratigraphy and telling the age of rocks by their characteristic fossils. Eventually, Smith could tell where he was in the sequence by the fossils alone, without even seeing the rocks from which they came. The gentlemen geologists who consulted him were amazed with his ability to correctly predict where the fossils in their collections had come from or to arrange their collections in stratigraphic order.

By 1799 Smith's lists of fossils characteristic of each formation were widely circulated among British geologists, but Smith was too busy working on the first geologic maps of England to publish this discovery for another decade or more. He produced a map of the rock units around Bath and Somerset in 1799, and by 1801 he had a rough map of many of the geologic units in England. This became the foundation for his most daring project, a complete map of the geology of all of England and Wales (figure 7.5). During the next few years, he worked as an independent mineral surveyor, studying and mapping all the rocks in the British landscape he could find and working for prominent people studying the geology of their large estates.

Meanwhile, other geologists were taking his discovery and making their own reputations from it, since Smith had not published and established his priority. In addition, Smith suffered from the prejudice of the wealthy, class-conscious gentleman geologists of the time, who viewed him as a lowly working man (as engineers and surveyors were then considered). Most of them considered geology a hobby, not a "vulgar" way of making a living. Smith was one of the few who might be considered a "professional" geologist in that he worked on geological problems for an income.

In addition, the idea of faunal succession was so powerful that it soon emerged in France. Baron Georges Cuvier and Alexandre Brongniart mapped the strata of the Paris Basin and eventually recognized their own sequence of French rocks and fossils. Some say that Brongniart may have heard of Smith's ideas during a visit to England in 1806, although the

Figure 7.4 ▶

Smith based his discoveries on careful observation and identification of the fossils in each layer: (*A*) An illustration from Smith's publication and (*B*) photos of some of Smith's fossils, now preserved in museum collections. (Courtesy of Wikimedia Commons)

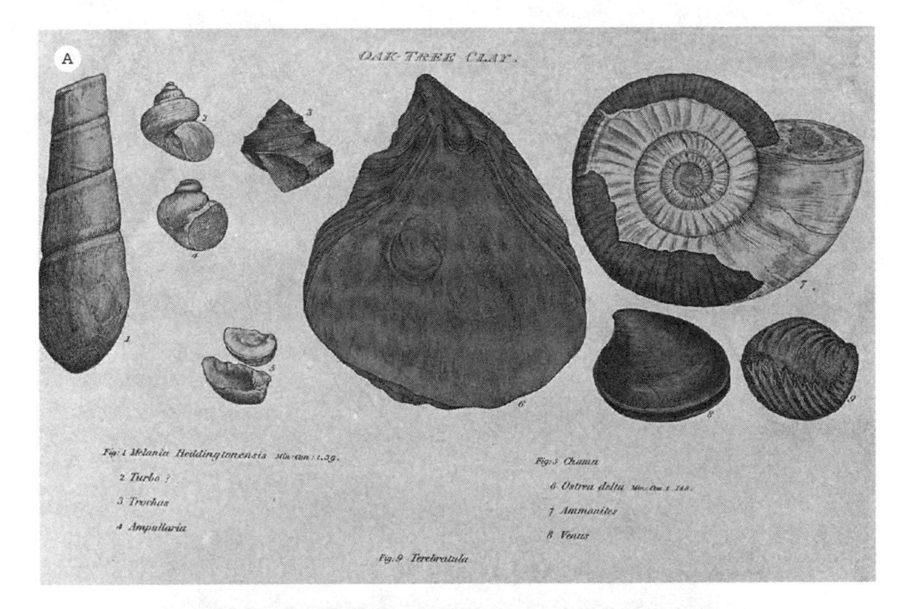

A
OAK-TREE CLAY.
1
2
3
4
5
6
7.
8
9
Fig: 1 Melania Heddingtonensis Min: Con: 1.39.
2 Turbo ?
3 Trochus
4 Ampullaria
Fig: 5 Chama
6 Ostrea delta Min: Con: 1. 148.
7 Ammonites
8 Venus
Fig. 9 Terebratula

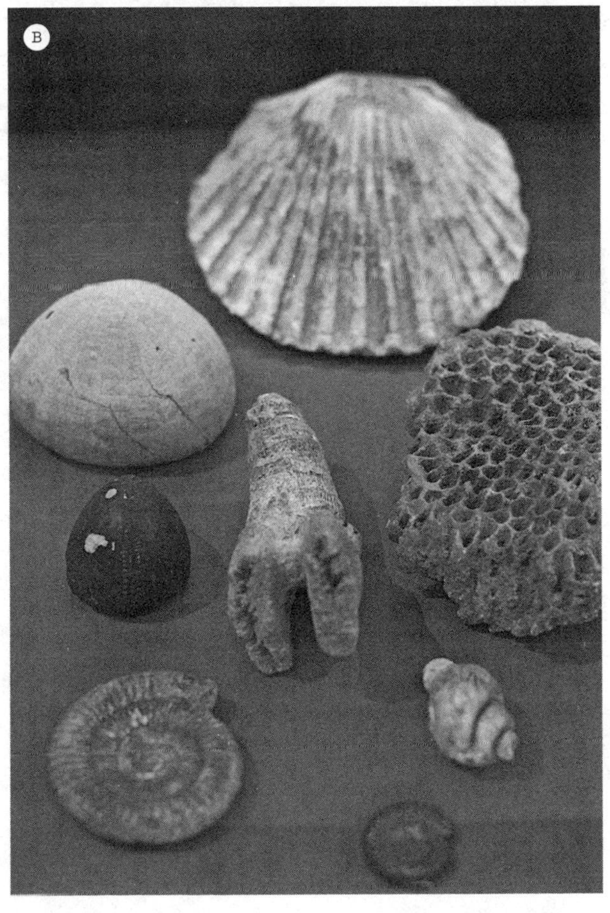

B

Figure 7.5 ▲
The geologic map of England, Wales, and parts of southern Scotland, as mapped, drafted, and published by Smith in 1815. (Courtesy of Wikimedia Commons)

French insist that Cuvier and Brongniart came up with the idea on their own. Whatever the truth, the idea of faunal succession was definitely in the air at the time.

PRISON AND VINDICATION

Smith poured most of his earnings into travelling and mapping all over Britain. Eventually, he published the first geologic map of England (1815), which is so good at its scale that it can still be used today (figure 7.5). Geologist Simon Winchester called it "the map that changed the world," because it launched the modern field of geology as a study of the three-dimensional arrangement of rocks in time and space and what this can tell us about earth history, the building of mountain belts, the movements of ancient seas, and especially the distribution of important mineral deposits. Every geologist since then has learned to make geologic maps early in his or her career, because it is one of the most essential tools in earth science.

In 1817 Smith took his map and drew a remarkable cross section all the way across the southern tier of England and Wales. It stretched from the ancient rocks of Mount Snowdon in northern Wales at one end, to the very young Eocene rocks beneath London on the other (figure 7.6). This was the first time that someone had drawn a geologic cross section on such a large scale and shown that the sequence of rocks across England had a regular, repeatable pattern that was easy to understand and represent by a tilted package of rock layers. Today, we point to places like the Grand Canyon to show this idea, but Smith was the first to demonstrate it, and without the benefit of good barren rock exposures that can be seen in a single vista. Although Smith himself never engaged in philosophical speculations about what it meant, it was the basis for the classic geologic column and time scale that is now so familiar to all of us (figure 7.7).

Other scholars of that time, such as Georges Cuvier and Alcide D'Orbigny in France, tried to reconcile this complex pattern of dozens of different faunal levels through time with the actual rock record, but the idea that all the world's rocks could be explained by a single Creation event and a single Noah's flood was rapidly being undermined. D'Orbigny went so far as to suggest that there had been 29 separate creation events and floods not described in the Bible, but clearly the old notions of Wernerian flood

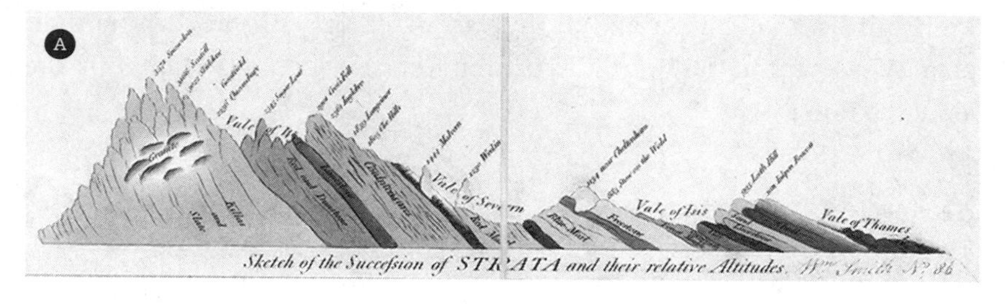

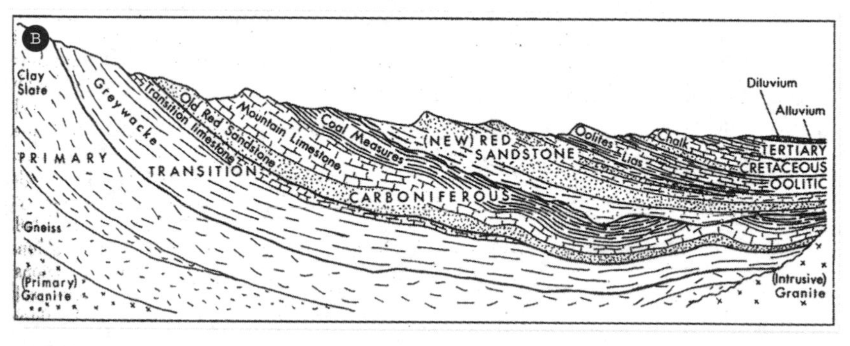

Figure 7.6 ▲

(*A*) Smith's 1815 east–west cross section through Wales to London, from Mount Snowdon and the oldest rocks in western Wales to the young rocks in the London area. (*B*) A modern version of the same cross section, with the units recognized by Smith in 1815. (Courtesy of Wikimedia Commons)

geology were being demolished as geologists actually mapped and documented the real sequence of rocks for the first time.

Unfortunately, other people had no qualms about plagiarizing Smith's work and publishing cheap knockoffs of his detailed map, so Smith made almost no money from his grand project. He eventually fell into debt when no one would buy his maps and charts and was sent to debtor's prison in 1819. When he was finally released, he found that he had lost his home of 14 years and all his property had been seized. He scratched out a living as an itinerant surveyor until one of the men who had employed him back in his better years, Sir John Johnstone, found out about his plight. Johnstone gave him full-time work in his estates in Scarborough in Yorkshire, where Smith not only improved his mapping of eastern England but developed the Rotunda, a museum dedicated to the geology of the Yorkshire coast. It is the oldest purpose-built museum in Britain. It was built as a tall, round tower, according to Smith's design suggestion, and the original display of

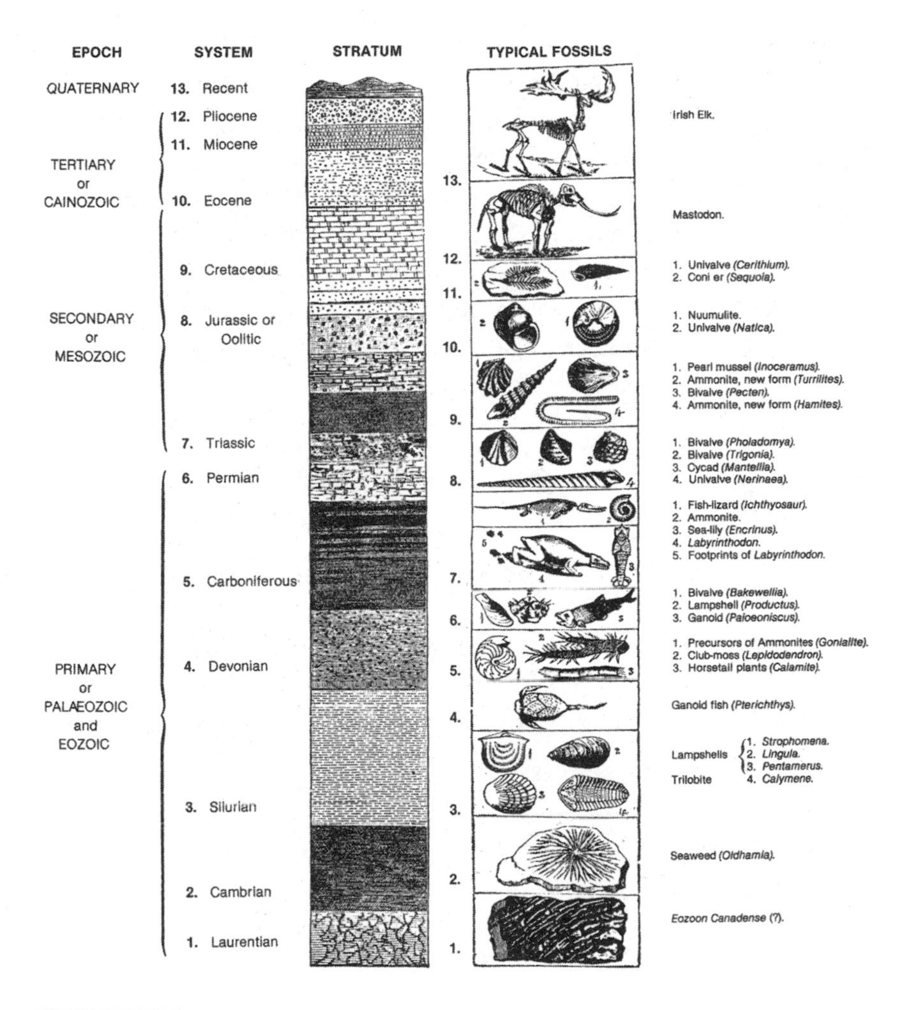

Figure 7.7 ▲

A diagram showing the known sequence of fossils as they were arranged by the 1840s, building on Smith's principle of faunal succession. (Courtesy of Wikimedia Commons)

fossils illustrated his ideas. The fossils and rocks were arranged in the order in which they occurred, with the youngest in the cases at the top and the oldest at the bottom. The order around the walls reflected the order of rocks on the Yorkshire coast. The museum is still popular today, although it has been modernized and upgraded in the past 180 years.

In 1831, toward the end of his life, Smith was finally given credit for all his groundbreaking work when was awarded the first Wollaston Medal and acknowledged by the Geological Society of London as the "father of English geology." In 1835 he was given an honorary doctorate by Trinity College in Dublin. And in 1838 he was appointed one of the commissioners who would select the site for the new Palace of Westminster. He died at age 70 in 1839 and was buried in Northampton near where he died. His grave is still visible in the churchyard of Saint Peter's Church in Marefair.

More importantly, his discovery of the principle of faunal succession was the foundation for telling geologic time, and Smith also drew the first geological column for an entire region. His development of the geologic map became the fundamental tool of all geology since then, literally, "the map that changed the world."

FOR FURTHER READING

Berry, William B. N. *The Growth of a Prehistoric Time Scale*. San Francisco: Freeman, 1968.

Rudwick, Martin J. S. *Earth's Deep History: How It Was Discovered and Why It Matters*. Chicago: University of Chicago Press, 2014.

Winchester, Simon. *The Map That Changed the World: William Smith and the Birth of Modern Geology*. New York: HarperCollins, 2001.

08 | CLOCKS IN ROCKS: ARTHUR HOLMES AND THE AGE OF THE EARTH

RADIOACTIVE URANIUM

[The concept of geologic time] makes you schizophrenic. The two time scales–the one human and emotional, the other geologic–are so disparate. But a sense of geologic time is the important thing to get across to the non-geologist: the slow rate of geologic processes–centimeters per year–with huge effects if continued for enough years. A million years is a small number on the geologic time scale, while human experience is truly fleeting–all human experience, from its beginning, not just one lifetime. Only occasionally do the two time scales coincide.

—ELDRIDGE MOORES, AS QUOTED BY JOHN MCPHEE IN *ASSEMBLING CALIFORNIA*

It is perhaps a little indelicate to ask of our Mother Earth her age, but Science acknowledges no shame and from time to time has boldly attempted to wrest from her a secret which is proverbially well guarded.

—ARTHUR HOLMES, *THE AGE OF THE EARTH*

IMPASSE

James Hutton wrote that the earth had "no vestige of a beginning," and after the first volume of Charles Lyell's *Principles of Geology* was published in 1830 almost all geologists agreed that the earth was immensely old. But just how old? How could we put a number on the age of the earth?

The problem was a hard one, but it failed to daunt scientists, who tried all sorts of ingenious solutions. The most common method was to add up the maximum thicknesses of all the sedimentary rocks on Earth, calculate how long it would have taken to deposit those layers, and use that as a minimum age. For example, they compiled the data for the maximum thickness of the Cambrian anywhere on earth, the maximum thickness of the Ordovician, and so on, then estimated how long the Cambrian or Ordovician might have lasted given typical rates of sediment accumulation. Most such estimates came to about 100 million years since the Cambrian, which we now know is off by a factor of almost 50. Why? As in all these early methods, there was a faulty assumption built into the calculation. The biggest factor was that they did not account for erosional gaps in the record, known as unconformities, when there is no rock representing a time interval. Later studies have shown that the rock record is full of time gaps and is actually "more gaps than record." Some geologists at the time suspected that there might a problem with unconformities, but no one could know how big a problem it really was.

Then there was the famous estimate by Irish physicist John Joly. He tried to calculate how long it would take for the oceans to go from freshwater to their current salinity, knowing the rate at which salt enters the oceans from the world's rivers. He also came to estimates of 80–100 million years, which we now know to be off by more than a factor of 50. What went wrong? Again, the problem was his faulty assumptions: that the oceans have been constantly adding salt since they formed. It turns out that the salt content of the oceans does not change much through time, because much of it gets locked into salt deposits in the earth's crust and seawater salinity is in equilibrium and stays very stable.

But the most famous and influential estimate was made by the renowned physicist William Thomson (later known by his title, Lord Kelvin). Kelvin made huge discoveries in the field of physics, especially in thermodynamics. The Kelvin temperature scale is named after him, since he pioneered the concept of absolute zero temperature (now known as 0 K—in case you were wondering, the unit of temperature is the kelvin, not the "degree" Kelvin, so there is no degree symbol with this unit). He was also a great inventor and helped create the transatlantic cable system that enabled telegraph and then telephone communications between Europe and North America. Thus, he was a giant among scientists of his time, and few people dared disagree with him.

In 1862 Kelvin attacked the problem of the age of the earth using thermodynamics. He assumed that the earth had started as a molten ball at the same temperature as the sun and that it had cooled off at rates that we can measure from the heat coming up from the earth's interior.

From this method, he estimated the earth was only 20 million years old, much younger than most geologists were willing to accept. It was also a problem for Charles Darwin, who knew that the earth had to be immensely old for his newly proposed concept of evolution to work. Kelvin's estimate didn't seem to offer enough time.

Through the rest of the nineteenth century, physicists and geologists were at an impasse. Neither group could comprehend the arguments of the other side or see the flaws in its own estimates. By the late 1800s, geologists began to give in and fudge their estimates a bit from the original values of 80–100 million years to numbers closer to Kelvin's 20 million years. Physics envy was just as powerful then as it is now! But the problem with Kelvin's estimate was just like the others: faulty assumptions. Kelvin made his calculation of the cooling of the earth by assuming that the heat was from the original solar system and no other heat sources were involved. We now know this is wrong. The earth does have an additional heat source.

IT'S RADIOACTIVE!

In 1896 Henri Becquerel in France discovered radioactivity, and in 1903 Marie and Pierre Curie showed that radioactive materials like radium produced a lot of heat. At the same time, New Zealander Ernest Rutherford was England's foremost authority on this new source of energy. In 1904 he was getting ready to address the Royal Institution of Great Britain about this new discovery, when he suddenly realized that the 80-year-old Lord Kelvin was in the audience. The young Rutherford was about to challenge the world's most famous physicist's estimate of the age of the earth! As Rutherford wrote later:

> I came into the room which was half-dark and presently spotted Lord Kelvin in the audience, and realised that I was in for trouble at the last part of my speech dealing with the age of the Earth, where my views conflicted with his . . . To my relief, Kelvin fell fast asleep, but as I came to the important point, I saw the old bird sit up, open an eye and cock a baleful glance at me.Then a sudden inspiration came, and I said Lord Kelvin had limited the age of the Earth, provided no new source [of heat] was discovered. That prophetic utterance referred to what we are now considering tonight, radium! Behold! The old boy beamed upon me.

Kelvin's estimate had been based on the faulty assumption that there were no other sources of heat beyond the earth's original heat when it cooled from a molten mass and that it would cool in no more than 20 million years. But radioactivity provides that additional heat. In fact, radioactivity provides so much heat that it is now the only source of heat that we measure coming from the earth's interior. The original heat from the cooling of the earth Kelvin thought he was measuring dissipated billions of years ago, maybe even during the 20 million years since the earth first formed 4.6 billion years ago.

GEOLOGIC TIME

Becquerel, the Curies, and Rutherford had pioneered the physics and chemistry of radioactivity, and their discoveries showed that Kelvin's assumption of no additional heat was wrong. But they were not geologists interested in determining the earth's age. It was two other scientists, Bertram Boltwood and Arthur Holmes, who realized that radioactivity provided a solution not only to Kelvin's dilemma of where the heat came from, but also the answer to the question: How old is the earth?

The method is a simple one, but widely misunderstood. There are just a handful of elements in nature that are radioactive and spontaneously decay from a parent atom (such as uranium-238, uranium-235, and potassium-40) to a corresponding stable daughter atom (lead-207, lead-206, and argon-40, respectively) at rates slow enough to use in geologic dating. This rate of decay is precisely known, so if we can measure the amount of both the parent atom and the daughter atom in a sample, their ratio is a measure of how long that decay has been ticking away.

Of course, there are lots of complications with real rocks, so very special conditions have to be met. The decay is measured from the time the decaying parent atoms are first locked into a crystal, so it primarily works on igneous rocks that cool from a magma, like lava flows, volcanic ash layers, and intruding bodies of magma forming dikes. Geochronologists (specialists in radiometric dating) try to get the freshest crystals possible to ensure there is no leakage and contamination of parent or daughter atoms leaving or entering the crystal that might distort the ratio. There are all sorts of laboratory procedures designed to eliminate expected problems in advance and to correctly calibrate the instrument used (known as a mass spectrometer, since it separates and measures the different isotopes by mass) to give reliable ages. Finally, every radiometric date comes with an error estimate based on how reproducible each result from the machine can be. So, for example, if they cite an age of 100 millions of years ago ± 5 million years (the unit for millions of years ago is Ma), they are saying there is a 95 percent probability that the true age lies somewhere between 95 Ma and 105 Ma.

But none of this was known in 1900 when radioactivity was just beginning to be understood. Physicists had been trying to date rocks by measuring the helium given off by uranium decay, but it was almost impossible to capture all the helium gas. Instead, it was Yale chemist Bertram Boltwood who discovered that uranium decayed to lead through radioactive breakdown. Following a suggestion from Rutherford, Boltwood noticed that the rocks he knew to be older had more lead in them than those that he knew to be younger. Unfortunately, he was using the very primitive understanding of the uranium-lead system that prevailed at the time. He didn't realize, for example, that there are two different radioactive isotopes of uranium, uranium-238 and uranium-235, each with a different decay rate and a different daughter isotope of lead. Nonetheless, he analyzed the samples he had, and in 1907 he obtained dates on samples ranging from 400 million years to as old as 2.2 billion years. This was the first evidence that the earth was indeed billions of years old, as geologists had long suspected, and that Kelvin's estimate was way off. Unfortunately, in his later life Boltwood suffered from severe depression, and his research came to a standstill. He committed suicide in 1927.

THE DATING GAME

Boltwood had done the first analyses, and his data indicated the ages of some rocks to be as old as 2.2 billion years, but he never followed up on his breakthrough. Thus, it fell to a young British geologist, Arthur Holmes (figure 8.1), to take the budding young field of geochronology and develop it into a rigorous science. Born to a family of modest means in 1890 in the tiny town of Gateshead (near Durham and the Scottish border), Holmes was originally planning to be a physics major at the Royal College of Science (now University College London). But in his second year, against the advice of his tutors, he took a course in geology and found his true calling.

Figure 8.1 ▲
A young Arthur Holmes in 1912, as he was beginning his career in geochronology and finishing graduate school. (Courtesy of Wikimedia Commons)

Holmes proved to be a brilliant undergraduate and was soon doing research. He latched onto the hot problem of radioactivity and realized that Boltwood's 1907 paper on uranium-lead dating held immense promise. For his undergraduate research project, he had a granitic rock from the Devonian of Norway to analyze. He cut his Christmas holiday short and spent his "holiday break" in London, working alone in the silent lab. As his advisor, physicist Robert Strutt, later recalled,

> We are at present largely subsisting on loaned apparatus, some of which belongs to other public bodies, such as the Royal Observatory, the Royal Society, etc., while some has been borrowed from private friends. I need hardly say that it seems rather below the dignity of an institution like the Imperial College that its teachers should have to beg apparatus of their personal friends for the purpose of teaching the students.

Holmes worked away in the cold, silent, lonely lab in January 1910, crushing the rock into a powder of mineral grains in a mortar made of agate, fusing the mineral grains in a platinum crucible with borax, dissolving it in extremely caustic hydrofluoric acid (see chapter 11), then boiling it again and again while measuring the radon emissions (an indirect measure of how much uranium was present). The lead content was measured by fusing the powder in a cake, then boiling it, dissolving it in hydrochloric acid twice, then letting it evaporate all its water. Then he heated it in ammonium sulfide to make the lead precipitate out as lead sulfide (known as the mineral galena). The precipitate was collected on a filter, dried, ignited, treated with nitric acid, boiled, treated with sulfuric acid, and heated again. Eventually, as Holmes wrote, "A tiny white precipitate then remained. This was collected on a very small filter, washed with alcohol, dried, ignited, and weighed with the greatest possible accuracy." Often there were only a few milligrams of material left.

These complicated chemical operations took incredible patience, extraordinary dexterity, and lots of time and often used up nearly all his original sample. Then to top it off, the results had to be verified, so the entire analysis was repeated two to five times, depending on how much original sample he had left. Once he had to discard all his data because radon had been leaking into the room. In other cases, he had to go begging at the British Museum for more sample, because he had used up his original allotment. Eventually, however, all this hard work paid off, and he got a reliable date of 370 Ma on his Devonian granite from Norway. He had greatly improved Boltwood's original methods and proven that uranium-lead dating could work and that it was possible to generate dates on rocks. The results were published in 1911, soon after he graduated in 1910.

Holmes was so poor after spending years living on a tiny scholarship of £60 a year that he quit school for a while and took a job prospecting for minerals in Mozambique to earn some real money. He spent 6 months there, found nothing, and suffered so extremely from malaria that his colleagues sent a letter home telling his family he was dead. He eventually recovered and managed to get a boat home, where he became a demonstrator (a low-level instructor) at his alma mater, Imperial College London. There he resumed his studies of the uranium-lead dating technique, figuring out that there were two different isotopes of uranium and lead, and this had to be understood in the dating analysis.

By 1913 he had so many new results and so many improvements on the method that he was able to write his groundbreaking book, *The Age of the Earth*, while he was still a graduate student. In it, he not only explained the basic principles of geochronology and discussed the problems with earlier methods of dating the earth, but also finally laid Lord Kelvin's mistaken estimate to rest. He had dates on some of the oldest rocks in Britain of 1.6 billion years old, although he refused to speculate on the age of the earth. Later editions included results from analyses of older and older samples, until by the 1950s, he had dates of 4.6 billion years, which is our present estimate. His early research earned him his doctorate from University College London in 1917. But World War I was raging in Europe, and it was hard to make a living on the paltry salary of a demonstrator. To make money for his family, he decided to try a career as an exploration geologist again, this time for an oil company in Burma in 1920. However, the oil company went bankrupt, and Holmes returned to England in 1924 flat broke. This was not his only tragedy: his 3-year-old son had contracted dysentery shortly after arriving in Burma and died there.

Luckily, after his return in 1924, his earlier reputation and research landed him a job as a "reader" in geology at Durham University, close to his birthplace. There he spent the next 18 years teaching geology and adding to and refining the database of radiometric dates from around the world. His work so dominated the field that he became known as the "father of geochronology" or the "father of the geologic time scale." In 1943 he moved north of the border to the University of Edinburgh, where he spent the last 13 years of his career until his retirement at age 66 in 1956.

A PIONEER OF PLATE TECTONICS

Meanwhile, through all those years teaching geology at the college level, Holmes accumulated the experience to write an introductory geology textbook. First appearing in 1944, Holmes's *Principles of Physical Geology* became the standard textbook for generations of British geology students and went through many editions. But this book was not entirely conventional. In the last chapter of his first edition, he fully embraced the controversial idea of continental drift espoused by German meteorologist Alfred Wegener back in 1915. This idea was roundly rejected by most geologists of the time, but Holmes had seen the evidence—rocks in Africa that matched those in South America.

Holmes went even further. Using his understanding of how radioactivity drove the heating of the earth's interior, he solved the mystery of Hutton's idea of "Earth's great heat engine." In a paper published in 1931 (figure 8.2), Holmes was the first to suggest that this heat drives great convection currents in the mantle that shift the continents above them. He even postulated the notion that the seafloor must spread apart, decades before any evidence of seafloor spreading emerged in the late 1950s.

Late in life, he began to receive honor after honor for almost single-handedly solving the problem of geologic dating. He was elected a fellow of the Royal Society in 1942 and won the Geological Society of London's Murchison Medal in 1940 and Wollaston Medal in 1946. He was given the Penrose Medal, the highest award of the Geological Society of America, in 1956. And in 1964, a year before he died, he received the Vetlesen Prize,

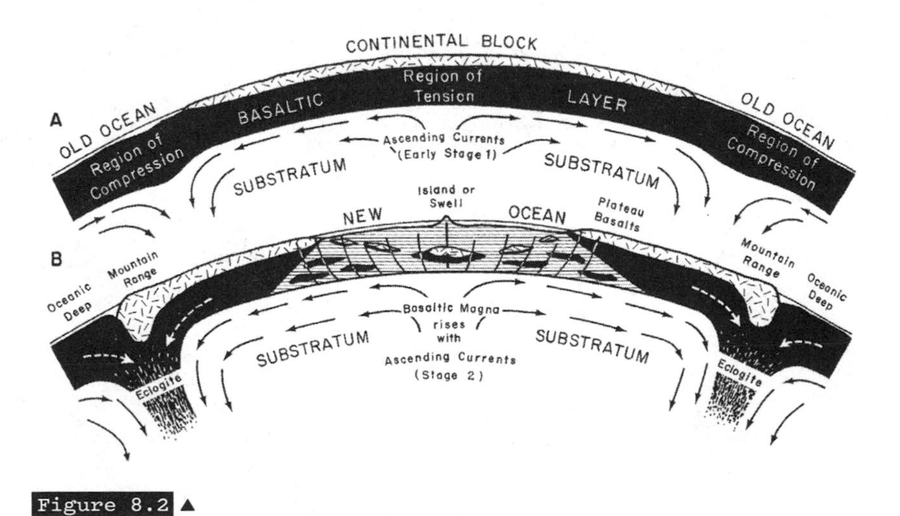

Figure 8.2 ▲
Holmes's diagram of mantle convection moving the continents around. From Arthur Holmes, *Principles of Physical Geology* (London: T. Nelson, 1944).

sometimes called the "Nobel Prize of Geology," for his contributions not only to geochronology but also to the newly ascendant field of plate tectonics.

FOR FURTHER READING

Dalrymple, G. Brent. *The Age of the Earth*. Stanford, Calif.: Stanford University Press, 1994.

Hedman, Matthew. *The Age of Everything: How Science Explores the Past*. Chicago: University of Chicago Press, 2007.

Holmes, Arthur. *The Age of the Earth*. London: Harper and Brothers, 1913.

Lewis, Cherry. *The Dating Game: One Man's Search for the Age of the Earth*. Cambridge: Cambridge University Press, 2002.

Macdougall, Doug. *Nature's Clocks: How Scientists Measure the Age of Almost Everything*. Berkeley: University of California Press, 2008.

09 | MESSENGERS FROM SPACE: THE ORIGIN OF THE SOLAR SYSTEM

CHONDRITIC METEORITES

Beyond matters of the soul, the inspiration for most art is in nature. For me, aesthetic meteorites are the closest approximation to being able to behold that which is in the heavens.

—DARRYL PITT

A BOLT FROM THE BLUE

It was a quiet night in the little town of Pueblito de Allende in Chihuahua, Mexico, on February 8, 1969. Everyone was fast asleep. Suddenly, at 1:05 in the morning, a huge fireball appeared out of the southwest and lit up the night sky and the ground brighter than daylight. It was a meteorite the size of a car, falling out of space at a speed of 10 miles per second (3,600 miles per hour). Residents were startled out of their sleep by the sounds of the flaming rock screaming down out of the sky and an immense explosion that rocked the earth as the meteorite hit. The impact scattered debris across an area spanning about 250 square kilometers, in a long oval 50 kilometers wide and 8 kilometers across, including thousands of small fragments of the meteorite itself, plus debris from the impact crater.

When dawn finally arrived and brought daylight to the landscape, the terrified residents emerged from their houses to see what had happened. They soon realized what had happened, as they picked up hundreds of pieces of the rock from space that had landed that night. Local officials and residents looked around for pieces of the meteorite, but luckily no one appeared to be injured or killed, and there was no serious damage.

Scientists responded as soon as the news spread and tried to reach the site. As recounted by Elbert King, a geologist at the University of Houston who specialized in meteorites, in his 1989 book *Moon Trip: A Personal Account of the Apollo Program and Its Science*:

> While unsuccessfully searching for a meteorite fall close to Crosby, Texas, I heard on the car radio about a very bright fireball witnessed in southern New Mexico, Texas, and northern Mexico. I returned to my office and asked my secretary, who was fluent in Spanish, to place some phone calls for me. I first contacted a newspaper editor in Chihuahua City. We had a lengthy conversation about the phenomena accompanying the meteorite fall but no specimens had fallen near Chihuahua City. Finally, I asked him the right question: "Do you know anyone who has any pieces of the meteorite?" "Oh yes," he said, and suggested that I call the newspaper editor in Hidalgo del Parral, much further to the south. My secretary located Sr. Ruben Rocha Chavez, editor of *Correa del Parral*. He recounted how a brilliant fireball had broken apart with a loud explosion in the middle of the night and had showered fragments over a large area near Parral. Chavez had several pieces of the meteorite on his desk and described them to me. There was no doubt—he had fragments of a freshly fallen stony meteorite! He invited me to visit Parral to see his pieces and to collect specimens. I thanked him for the information and his invitation and told him I would be there as soon as possible.
>
> A quick check of airline schedules showed it was not going to be easy to get to Parral. I could fly to El Paso, but that was still more than three hundred miles north of Parral. It was the fastest way, however. My secretary promised to cover me with paperwork. I stopped by my house for a few clothes and headed for the airport.
>
> The plane took off on time, but, as luck would have it, a faulty landing gear indicator light grounded us in San Antonio for five hours while it was replaced. By the time I arrived in El Paso it was already dark. I picked up a rental car, cleared through customs, and drove south. It was important to recover pieces of the meteorite right away in order to measure their short half-life radioactivities. This would be great practice for the Radiation Counting Laboratory of the LRL [Lunar Research Laboratory in Houston]. The Mexican highways were difficult to negotiate in the dark. The best technique was to follow a hundred yards behind a car with Mexican license plates. Some of the drivers were going 80 miles per hour, and when I saw brake lights or a cloud of dust, I knew

the driver had spotted a burro on the highway. I arrived in Parral just after dawn. I checked into a hotel, washed up, drank some strong coffee, ate eggs and tortillas, and went to look for the newspaper office. I was waiting when the editor arrived. I was astonished when I saw the two big meteorite pieces on the editor' s desk. One weighed more than 30 pounds.

The greatest surprise was the meteorite type—a rare carbonaceous chondrite. Chondrites are stony meteorites that contain chondrules, small spheres of silicate of disputed origin. Carbonaceous chondrites are chondrites that contain abundant carbon and organic compounds. While I was standing in Chavez's office, the telephone rang. The editor handed the receiver to me. It was a colleague from the Smithsonian who wanted information about the meteorite. He had called my Houston office, where my secretary gave him the number of the newspaper office. I told him what little I knew. I asked the editor about his plans for the two specimens on his desk. He said they were reserved for the National Museum. I agreed this was perfectly appropriate, but I was eager to recover some additional specimens. The editor said I must visit the local municipal president or mayor. I was going to be treated as an official NASA representative.

The mayor, Sr. Carlos Franco, was extremely gracious, and though my Spanish was meager and he spoke little English, we had an amiable meeting. I explained, through the editor as translator, how scientifically important meteorites are in general and that this particular one was a very rare type. Sr. Franco was eager to help me, and he assigned me one of his policemen and an official car for as long as I needed them.

We drove to places where specimens had been found. Recovering additional specimens proved to be easy. Everyone had small pieces of the meteorite, but I wanted some larger ones. I purchased these from the local people, with the policeman acting as interpreter and handling the negotiations. We documented several sites where specimens had been found. The stones had showered over a large area. One large stone had missed the post office in Pueblito de Allende by only 30 feet. Meteorites normally are named after the nearest post office. This one almost named itself. We listened to many tales of the fireball, its direction of travel, the loud claps of thunder, stones falling everywhere, and people running to the church in the middle of the night. I picked up 13 pieces of the meteorite, including two large ones—enough samples for the time being.

King was followed by many other scientists from museums and universities around the world, and soon the search was on for as many fragments of the Allende meteorite as could be found. It came at a particularly important time in the study of meteorites, because the field of planetary science was growing and well funded thanks to the Apollo missions. Two of the Apollo 11 crew members (Neil Armstrong and Buzz Aldrin) were the first humans to land on the moon, and their moon walk happened only a few months after the Allende impact. Altogether, thousands of pieces totaling over 3 metric tonnes were collected, and people are still finding small pieces today, some 48 years later. In fact, small fragments only a gram or two in weight are still sold over the Internet and are actually quite affordable. Thanks to the interest surrounding it, Allende is by far the best-studied meteorite in history.

TRACES OF THE EARLY SOLAR SYSTEM

The Allende meteorite was part of a class of meteorites known as carbonaceous chondrites, which are among the rarest and most important meteorites. Allende was very unusual in that most of the meteorites known to be in this class had been collected many years earlier and had been sitting in museum drawers. Prior to Allende, the best-studied of the carbonaceous chondrites was the Orgueil meteorite from France, which had landed back in 1864. A number of other smaller examples were known but poorly studied. All of the Orgueil meteorite's short-lived isotopes had long since decayed, and some of the meteorites were weathered and altered by having lain out in the field a long time before they were collected. By contrast, Allende was a fresh fall, so the material could be analyzed mere days after it had fallen, and there had been no time for weathering or contamination.

Chondritic meteorites are a special class of meteorites that formed from the early solar system and that predate the planets. They are made of the dust and debris of the original solar dust cloud, or solar nebula, or from smaller planetary bodies that never became large enough to have a separate core and mantle, so they are valuable clues to early solar system evolution. They get their name from the tiny blobs of material found in them (figure 9.1), called chondrules, which are even older bits of the early solar system that clumped together when the meteorite coalesced.

Although small chondrites are not rare (about 2,700 specimens in collections or about 86 percent of total meteorites are chondrites), certain types of chondrites are very rare. Among these are the carbonaceous chondrites like Allende, which make up less than 5 percent of the total number of chondrites.

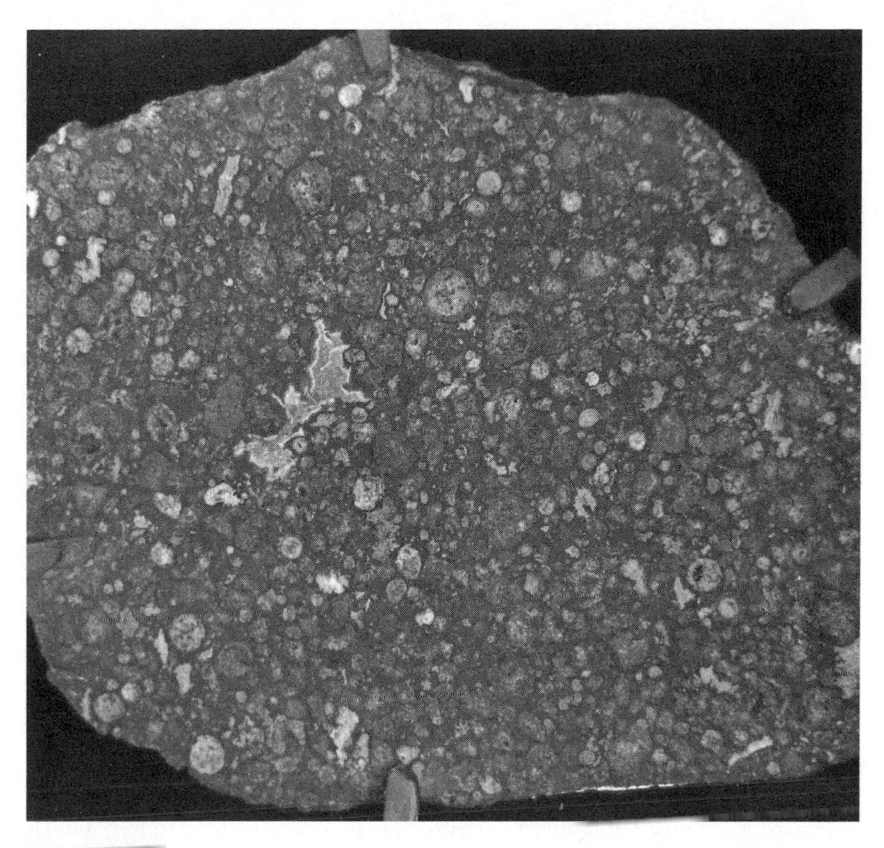

Figure 9.1 ▲
A slice of the Allende meteorite, showing the densely packed blobs of early solar system material known as "chondrules." (Courtesy of Wikimedia Commons)

They get their name because they have a relatively high carbon content compared with other meteorites and often still contain water-bearing compounds from the primordial solar system. It is thought that this is because they formed farther away from the sun than other meteorites and were not heated enough to wipe out their carbon or water content.

Once scientists had samples of Allende in their labs, they mined those remnants for every possible bit of information. The overall composition of the matrix surrounding the chondrules gave a good idea about the composition of the initial dust ring of the solar system. Other scientists focused on the chemistry of the chondrules included within the meteorite. The most interesting of these are known as "CAIs," or calcium-aluminum-rich inclusions. They have a truly unusual composition, rich in not only calcium and aluminum, but also silicon, oxygen, iron, and other elements. Their composition is completely unlike the rest of the early solar system, so they are thought to have formed from a high-temperature (greater than 1,300 K) protoplanetary disk of matter in the earliest stage of the solar system, before most of the rest of the material had condensed out.

In addition to giving us insights into the earliest history of the solar system, carbonaceous chondrites also give us dates for the time when the solar system formed. Allende has chondrules (including CAIs) that gave a uranium-lead date of 4.567 billion years in age. This is 30 million years older than the formation of Earth, and about 200 million years older than the oldest rocks and minerals on Earth. Another CAI in a different carbonaceous chondrite from northwest Africa gave an age of 4.56822 ± 0.00017 billion years, making it the oldest object ever dated, and a good estimate for the beginning of solar system formation.

Studies of the Allende meteorite and other carbonaceous chondrites continue to be published as scientists think of new things to analyze and newer and better techniques come along that weren't available in 1969. In 1971, scientists discovered tiny black markings (up to 10 trillion per square centimeter) that were evidence of radiation damage. This proves that the meteorite did not start out near Earth (which is screened from radiation by its magnetic field), but formed far from Earth and before it had a magnetic field, when objects (including the oldest lunar rocks) were being intensely bombarded by radiation. Using the same Caltech lab that first analyzed the lunar rocks (waggishly nicknamed the "Lunatic Asylum"), scientists in 1977 found that the Allende meteorite contained new forms of the elements calcium, barium, and neodymium, as well as krypton, xenon, nitrogen, and other rarer elements that apparently came from the shockwave of a supernova that may have helped trigger the formation of the solar system.

Even more importantly, the Allende meteorite was rich in the rare isotope magnesium-26, which is formed from the decay of radioactive aluminum-26. This decay takes place very rapidly and must have occurred soon after the solar system had formed. But the abundance of magnesium-26 in this meteorite suggested that it was once abundant in all solar system rocks, including those that formed the earliest earth, and answered a long-standing question: What heated the early earth and caused it to melt, separating its core from its mantle? The answer? Abundant aluminum-26 in the early earth generated more than enough heat by its decay to melt the earth many times over.

LIFE IN METEORITES?

In fact, 1969 turned out to be a banner year for meteorite studies. On September 28, 1969, another carbonaceous chondrite (figure 9.2) fell near Murchison, in Victoria, Australia. Local people first saw a fireball at about 10:58 A.M., followed by the sound of its descent through the atmosphere, and then felt its impact, a tremor that occurred about 30 seconds after

Figure 9.2 ▲
One of the larger pieces of the Murchison meteorite, now on display at the National Museum of Natural History, Smithsonian Institution. (Courtesy of Wikimedia Commons)

the fireball was sighted. The meteorite broke into three large pieces as it fell, then broke up even more after impact, forming a strewn field with an area of more than 13 square kilometers. Hundreds of fragments were found, totaling over 1,000 kilograms in weight. Many weighed more than 7 kilograms, and the largest weighed 680 kilograms and broke through the roof of a barn, landing in a pile of hay.

The Murchison meteorite turned out to be even more important than most other carbonaceous chondrites, because it contains organic compounds not found in any previous meteorite. The original studies found 15 amino acids, and more recent research with more sensitive techniques has found up to 70 amino acids and many more complex compounds. The discovery of amino acids was a shock, because amino acids are the building blocks of life, and it was thought that they could only be produced in a warm little pond on Earth. Back in 1953, the famous Miller-Urey experiment simulated the early earth's atmosphere and ocean in a lab apparatus. Stanley Miller and Nobel Prize–winning chemist Harold Urey showed that simply by heating a mixture of ammonia, methane, nitrogen, and water (but no free oxygen), the early earth could have produced most of the amino acids used by life. Now the Murchison meteorite showed that the process must have been widespread and had actually happened, and that amino acids were formed all over the early solar system, long before the earth formed. In fact, many scientists argued that life on Earth was seeded from amino acids that rained down from space, so in a sense, all life is extraterrestrial in origin.

Even more importantly, the amino acids in the Murchison meteorite were a mix of both right-handed and left-handed forms. This a property of chemical compounds in which molecules are asymmetrical and one form is the mirror image of the other. This showed that even if life started on Earth from a shower of amino acids brought by meteorites (or if it formed by itself in a warm little pond on Earth), there is only one common ancestor to all of life, because all biologically important molecules (except certain sugars) are left-handed—a property they must have inherited from a single early living form that happened to use only left-handed molecules.

So the next time you visit a museum and see a carbonaceous chondrite on display (especially if it's a piece of the Allende or Murchison meteorite), show it some respect. It is probably the oldest object you will ever encounter, and it's a piece of the earliest solar system before the planets formed.

Even more, it may have carried the seeds that launched the origin of life on Earth in the first place.

FOR FURTHER READING

Bevan, Alex, and John De Laeter. *Meteorites: A Journey Through Space and Time.* Washington, D.C.: Smithsonian Books, 2002.

Chambers, John, and Jacqueline Mitton. *From Dust to Life: The Origin and Evolution of Our Solar System.* Princeton, N.J.: Princeton University Press, 2013.

Dalrymple, G. Brent. *The Age of the Earth.* Stanford, Calif.: Stanford University Press, 1994.

——. *Ancient Earth, Ancient Skies: The Age of the Earth and Its Cosmic Surroundings.* Stanford, Calif.: Stanford University Press, 2004.

Gargaud, Muriel, Hervé Martin, Purificacíon López-García, Thierry Montmerle, and Robert Pascal. *Young Sun, Early Earth, and the Origins of Life: Lessons for Astrobiology.* Berlin: Springer, 2013.

Hedman, Matthew. *The Age of Everything: How Science Explores the Past.* Chicago: University of Chicago Press, 2007.

Macdougall, Doug. *Nature's Clocks: How Scientists Measure the Age of Almost Everything.* Berkeley: University of California Press, 2008.

Nield, Ted. *The Falling Sky: The Science and History of Meteorites and Why We Should Learn to Love Them.* New York: Lyons, 2011.

Norton, O. Richard. *Rocks from Space: Meteorites and Meteorite Hunters.* Missoula, Mont.: Mountain Press, 1998.

Smith, Caroline, Sara Russell, and Gretchen Benedix. *Meteorites.* London: Firefly, 2010.

Zanda, Brigitte, and Monica Rotaru, eds. *Meteorites: Their Impact on Science and History.* Cambridge: Cambridge University Press, 2001.

10 | THE CORES OF OTHER PLANETS

IRON-NICKEL METEORITES

My dad took me out to see a meteor shower when I was a little kid, and it was scary for me because he woke me up in the middle of the night. My heart was beating; I didn't know what he wanted to do. He wouldn't tell me, and he put me in the car and we went off, and I saw all these people lying on blankets, looking up at the sky.

—STEVEN SPIELBERG

CRATER OF CONTROVERSY

In the middle of the Painted Desert in Winslow, Arizona, you can travel west on Interstate 40 just another 29 kilometers (18 miles), or east from Flagstaff on Interstate 40 just 60 kilometers (37 miles), and you come to the turnoff for Meteor Crater (figure 10.1). The signs suggest that it's just another tourist trap, but it's not. It's just one of the most amazing natural sights in the United States that's not under the protection of the National Park Service, National Forest Service, Bureau of Land Management, or any other government agency, but privately owned. It was originally known as the Canyon Diablo crater, after a ghost town about 19 kilometers (12 miles) northwest of it. Another old name for it was Coon Mountain Crater.

Early geologists, like the legendary Grove Karl Gilbert of the U.S. Geological Survey who published a paper on Meteor Crater in 1892, insisted it was a volcanic crater. This was not unreasonable, because volcanic craters are common in the area, especially in the area north of Flagstaff, where the San Francisco Peaks and Sunset Crater are all examples of very recent volcanism. Gilbert had a sterling reputation. He had mapped and described

Figure 10.1 ▲
Barringer's Meteor Crater: (*A*) Aerial view; (*B*) a view from the rim. (Courtesy of Wikimedia Commons)

many of the key features of the Colorado Plateau, had proved that giant lakes once flooded Salt Lake City and the Bonneville salt flats, and also had been on the scene to document the geologic changes caused by the San Francisco earthquake in 1906.

Gilbert had studied the crater intensively, and originally concluded that it was volcanic, or possibly a steam explosion, although at one time he had taken the meteorite impact idea seriously. According to Gilbert, the nearby scatter of meteorites was only a coincidence. His main evidence against an impact was the lack of meteoritic material within the crater. The volume of shattered rocks around the rim only filled the crater itself, and there was no extra mass of iron and nickel in the middle, nor a magnetic signature that a meteorite was buried deep in the crater. Thus, most geologists sided with Gilbert and doubted that the crater was anything other than a volcanic vent.

However, others disagreed with the prevailing view of the geologists. Among these was a mineralogist named Albert E. Foote. Several years before, Foote had received samples from a local railroad executive who had built the train line that now parallels Interstate 40 and realized at once that the samples were meteorites. He led an expedition to the crater, and found hundreds of fragments, including one weighing over 600 pounds. These meteorites (figure 10.2) were distinctive in that they were made mostly of iron and nickel, a class of meteorites with great significance. Some of them even had tiny diamonds in them, proof that they had experienced extreme pressures and temperatures. Foote carefully described the crater and published his findings in the journal of the American Association for the Advancement of Science, still one of the most prestigious scientific journals in the world.

Another man who disagreed with the volcanic explanation was mining engineer and businessman Daniel M. Barringer. Back in 1894, Barringer had made over $15 million on the Commonwealth silver mine in Cochise County, Arizona, so he was experienced with mining operations and had

Figure 10:2 ▲

A piece of the Canyon Diablo iron meteorite. (Courtesy of Wikimedia Commons)

the capital to invest. He was so sure that this was a meteor crater that he bet his own money on it. His Standard Iron Company bought the land in and around the crater and staked a mining claim in 1903. His land patent was signed by none other than President Theodore Roosevelt himself, and the president even authorized the founding of a post office at the crater rim—designated Meteor, Arizona—to serve the site. Barringer and his company studied the crater between 1903 and 1905 and found convincing evidence that it was indeed an impact crater. About 30 tons of large meteorite fragments had been discovered around the crater since the early days, so Barringer believed that the bulk of the meteorite must be beneath the crater floor. Based on his mining experience, he estimated that the meteorite had a mass of 100,000,000 tons, worth over a billion in 1903 dollars.

A monument to persistence, Barringer and his company drilled down below the bottom of the crater for 27 years, reaching a depth of 419 meters (1,375 feet), but never found any significant iron deposits. The company spent over $600,000 in exploration costs but gave up when Barringer died in 1929. Bitter and disappointed, Barringer never stopped believing he was right, but his failure further confirmed the beliefs of most geologists that it had been a volcanic crater all along. Ironically, astronomer Forest Ray Moulton performed calculations in 1929 that showed the meteorite should have been vaporized, and Barringer's quest would have been in vain from the beginning. Barringer apparently read Moulton's paper before he died.

It was not until the 1950s that planetary physics began to mature as a field, and scientists realized that most meteorites burn up in the atmosphere or vaporize upon impact, leaving only small amounts of the original material scattered around. This was first pointed out in 1930 by Herman Leroy Fairchild, an early promoter of impact cratering as an important process.

In the 1960s and 1970s, the impact origin of Meteor Crater (also known as Barringer Crater) was finally proven by Gene Shoemaker of Caltech and the U.S. Geological Survey, one of the pioneers of planetary geology. (I knew Gene during my days at Caltech and did a lot of my magnetic analyses in the lab where he worked.) In 1960, Shoemaker took samples from the crater and discovered minerals like coesite and stishovite, forms of quartz that are only formed by the shock of an impact (or a nuclear explosion, where these minerals were first detected). This was definitive proof that the crater had not been formed by a volcanic eruption. Later, Shoemaker carefully remapped the crater. He found that the beds of impact debris surrounding the rim are stacked in upside-down order, with the debris from the lowest unit in the crater (the Coconino Sandstone) sitting on the top of the rim deposits and debris from the youngest unit (the Moenkopi Shale and Kaibab Limestone) lying at the bottom of the rim deposits. This upside-down sequence could only be explained if the impact had blasted the layers overlying the explosion upward and then sideways in a large flap that then flopped over upside-down after it landed, then broke up.

Today, planetary scientists estimate that the impacting body was an iron-nickel meteorite that struck about 50,000 years ago, when Ice Age mammals like ground sloths and mammoths roamed across the wetter, more wooded late Ice Age landscape. No Paleoindian cultures had reached North America yet, so there were no human witnesses. The meteorite was originally about 50 meters (160 feet) across and weighed almost 300,000 tons, three times the size that Barringer had estimated. It flew out of the sky at a velocity of about 20 kilometers per second (12 miles per second, or 4,300 miles per hour) and hit the ground with the impact of a 10-megaton nuclear bomb. In fact, there are craters from nuclear bomb tests that are just this size, and they closely resemble Barringer Crater. About half of the meteorite's original mass was vaporized on impact, and the rest was scattered around the landscape, which is why so little was left in the crater itself for Barringer to find.

Fortunately, Barringer's property claims were passed on to his descendants, so the family now owns and operates the crater visitor center and museum, which sits on the north rim with a spectacular view across the crater. Barringer's heirs make far more money from tourists than Barringer would ever have earned through mining. Meteor Crater has been used to train NASA astronauts preparing for the moon walk, for meteorological experiments, and in numerous films as a setting of unusual scenic interest. And in 1982, the Meteoritical Society named its highest honor the Barringer Medal in honor of the prophet of impacts who was scorned by all his contemporary geologists and lost millions on his dream of finding the meteorite.

VISITORS FROM THE SKY

Huge bodies of iron falling from the sky are impressive to any culture, and some were worshipped as special objects sent down to the earth by their gods. For example, the Clackamas tribe of Oregon worshipped the

Willamette meteorite. Many scientists think the black stone in the corner of the Kaaba, the shrine in Mecca that is worshipped by millions of Muslim pilgrims each year, may be an iron-nickel meteorite. In other cases, iron-nickel meteorites served as a useful source of iron for prehistoric tools, but this became less important during the Iron Age, when smelting allowed people to make tools out of more readily abundant sources of iron.

Some of these objects are truly spectacular in size. The largest is the Hoba meteorite in Namibia, which has never been moved from its site because it weighs at least 60 tons (figure 10.3). It was discovered in 1920 when the landowner hit it with his plow; the meteorite was completely buried and any crater it might have formed has long since eroded away. It is thought to be so large because it apparently slowed down to terminal velocity as it approached the earth's atmosphere. It maybe even skipped a few times due to its flat shape and smooth flat faces, so it did not hit the ground hard enough to make a big crater or vaporize, as many meteorites do.

Figure 10.3 ▲

The Hoba meteorite in Namibia, the largest ever found. It is too large to be moved, so a monument has been built around it. (Courtesy of Wikimedia Commons)

It is now a national monument protected from vandalism and is visited by thousands of tourists each year.

Another huge object is the Cape York meteorite, which fell near Greenland's Cape York about 10,000 years ago. It broke into many pieces, and the Inuit people of north Greenland have been using them to make iron tools like knives and harpoons for centuries. The largest piece (figure 10.4A), known to the Inuit as *Ahnighito* ("the Tent" in Inuit) weighed 31 tons and measured 3.4 meters by 2.1 meters by 1.7 meters—larger than a small truck; another piece, "the Woman," weighed 3 tons; and a third piece, "the Dog," weighed 400 kg (880 pounds). Stories of these objects reached scientists as early as 1818, and five expeditions were mounted between 1818 and 1883 to locate the source of all the fragments of this meteorite. It eventually was found in 1894 by the famous explorer Robert Peary, who is believed to be the first to reach the North Pole (in 1910). (There's some controversy about whether Peary reached it first or not, as there was a rival claim from explorer Frederick Cook). It took 3 years to move Ahnighito and some other pieces to the coast, forcing Peary's crew to build a small railroad (the only railroad Greenland has ever had) to move them. Peary then sold it to the American Museum of Natural History in New York for $40,000 in 1897 dollars, and it is still on display there today. It is the heaviest meteorite ever moved, and its display stand has supports beneath it that reach to the bedrock under the museum to avoid putting stress directly on the floor of the building.

Also in the same museum (the American Museum of Natural History in New York) is the most famous of all iron-nickel meteorites, the Willamette meteorite (figure 10.5). It is the largest meteorite ever found in North America, and the sixth largest in the world. It was found near West Linn in the Willamette Valley of Oregon by the Clackamas people, who called it *Tomonowos* ("Visitor from the Sky"). The Willamette meteorite had no impact crater around it, because it was moved by glaciers that plucked it from Montana or Canada about 13,000 years ago. It weighs 15 metric tons (15,000 kilograms) and is about 3 meters (10 feet) long by 2 meters (6.6 feet) wide and 1.3 meters (4.25 feet) deep. Oregon settler Ellis Hughes "discovered" it in 1902 (ignoring the previous discovery by the Clackamas) and realized that it was on land of the Oregon Iron and Steel Company. He then secretly moved it over to his own land, covering 1,200 meters (0.75 miles) in 3 days, then filed a mining claim for it. But the Oregon Iron and Steel Company found out, sued him, and won possession in court in 1905.

Figure 10.4 ▲

Fragments of the enormous Cape York meteorite: (*A*) *Ahnighito*, "the Tent," the largest of the fragments. It is now on display in the American Museum of Natural History, New York; (*B*) *Agpalilik*, "the Man," on display at the University of Copenhagen Museum. (Courtesy of Wikimedia Commons)

Figure 10.5 ▲

The Willamette meteorite, now on display in the American Museum of Natural History, New York. (Photo by the author)

They then sold it to the widow of millionaire William E. Dodge for $26,000 (about $700,000 in today's dollars), who gave it to the American Museum of Natural History. There it stands on display, a huge object that impresses everyone who sees it, and more than 40 million people have seen it in the past 112 years, more than any other meteorite known. In 1999 the Grande

Ronde tribe sued to have it returned to them, and in a court settlement, the museum kept the meteorite, but the Clackamas have the rights to do a private ceremony around it once a year. If the museum ever removes it from display, it must go back to Oregon. Meanwhile, there is a replica outside the University of Oregon Museum of Natural and Cultural History in Eugene.

FRAGMENTS OF THE CORE

Iron-nickel meteorites are a very rare and special kind of meteorite. Only 6 percent of known meteorites are of this composition, since stony meteorites and chondrites are far more abundant. But an iron-nickel meteorites is impressively heavy when you pick one up and heft it. Since they are so much denser than stony meteorites or chondrites, they make up 90 percent of the mass of known meteorites. They are also overrepresented in collections, because they look distinctive (even to a layman), they are more resistant to weathering on the earth's surface, and they are also more resistant to ablation from entry through the atmosphere.

As their name suggests, these objects are made mostly of iron, with about 5–25 percent nickel and minor amounts of cobalt and other rarer elements. Thus, they are chemically much simpler than the stony meteorites and chondrites, which have many different chemicals and minerals in them.

But the most interesting aspect of iron-nickel meteorites is that they provide samples of what makes up the core of many planets (including ours). When the spectra of certain asteroids (known as M-type) are analyzed, they turn out to be the same composition as iron-nickel meteorites. From the geochemical evidence trapped inside them, we know that iron-nickel meteorites originally formed the cores of large protoplanets that have since broken up. They also trap isotopes of magnesium-26, which as we discussed in chapter 9, was the radioactive heat source that melted protoplanets and allowed their denser materials (iron and nickel) to sink to their cores and separate from their mantles during planetary differentiation.

This information is consistent with what geophysical evidence tells us about the core of the earth. Seismology tells us about the size of the core (2,900 kilometers beneath our feet at the bottom of the mantle). Seismology plus gravity measurements suggest that the core has a density about 10–12 times as dense as water, which could only be accomplished by a very dense metal under enormous pressure. Finally, the fact that the earth has a magnetic field suggests that the core must be a good electrical conductor, which suggests metals like iron and nickel. Thanks to these meteorites, we know that the only common materials in the original solar system that fit all these properties (density, electrical conductivity) are iron and nickel, so the only reasonable explanation is that the earth has an iron-nickel core as well.

LEAD, LEAD, EVERYWHERE

But how old are these meteorites, such as the Canyon Diablo meteorite from Barringer Crater? The lack of anything but iron and nickel and a few other metals like cobalt and lead in them (compared with the silicate minerals in stony meteorites and chondrites) ruled out a lot of the traditional dating techniques, such as potassium-argon dating or rubidium-strontium dating. Even conventional uranium-lead dating was problematic, since the rocks were 4 billion years old and older, so almost no measurable amounts of the parent atom uranium were left. What technique might work with such old objects?

The problem fell to a young chemistry student, Clair "Pat" Patterson, in 1948 (figure 10.6). Born in Mitchellville, Iowa, in 1922, Patterson went to Grinnell College in Iowa, then got his master's at the University of Iowa working on mass spectroscopy. During the war, both he and his wife

Figure 10.6 ▲
Clair "Pat" Patterson at Caltech. (Courtesy of the Caltech Archive)

(also a chemist; they met at Grinnell) were recruited to work on the Manhattan Project to develop the atomic bomb. Once the war ended, Patterson began his Ph.D. work at the University of Chicago. His advisor, Harrison Brown, had an insight for a method of dating by measuring the lead daughter products (lead-206 and lead-207) produced by the decay of the two different isotopes of uranium, uranium-235 and uranium-238. Since the uranium decays to lead at different rates in each system, the ratio of the two could be put on a plot that would give a slope that depended on the age of the samples. Another University of Chicago student, George Tilton, would work on a parallel problem with uranium counts as a cross-check on their results.

It seemed simple in theory, so Patterson set about trying to make his measurements. To his dismay, his results were scattered all over the place and were clearly too noisy. Something was adding lead to the system; he was measuring more background lead contamination than lead in the sample. Patterson tried to remove all contaminants from his lab. In the "clean room," researchers had to first take a shower and then wear special protective clothing to keep contaminants from their clothes from spreading, cover their feet with booties and their heads with hair covers, and wear surgical masks. Every surface of the room and the machinery was cleaned, top to bottom. Over and over they worked at removing every possible way that lead from outside could contaminate their lab work. Once he had made the room super clean, he started to get good results. In 1953 he showed that the Canyon Diablo meteorite was 4.54 ± 0.05 billion years old, the oldest object ever dated to that point, so the age of the earth's core (and probably the solar system) was also 4.5 billion years old.

Meanwhile, Patterson had been appointed to set up the geochemistry program at Caltech, and he built an even better clean lab there. (He remained at Caltech until he retired, and I got to see his old clean lab when I was at Caltech.) After all his work, Patterson wondered if there was lead contamination in everything, even the air from outside of the lab. His machines were so sensitive by this point that he was able to measure tiny amounts of lead in the air, the water, and in many other substances. To his surprise and horror, it was apparent that there was lead in nearly everything, especially in our bodies, from ingesting lead in our water and food. By looking at samples of water from ice cores from Greenland, he could establish that this lead contamination was very recent. In fact, the lead starting showing up at the same time that oil companies began to put lead in gasoline to reduce knocking and pinging in engines. Lead was also used in paint, glazes, food containers, and even water distribution systems. People have known that lead was toxic for over a century. Many scholars think one of the things that destroyed the Roman Empire was lead poisoning from the pipes that carried their drinking water. Yet somehow no one thought that putting lead in so many products might potentially contaminate the environment.

Patterson finally published his results in 1965, and immediately encountered backlash from the powerful industries who were poisoning us, including oil companies, lead mining companies, the lead additive industry, and especially their lobbying groups. The Ethyl Corporation (a lobbying group that advocated the use of tetraethyl lead in gasoline) hit him with everything they had. Their own expert chemist, Robert Kehoe, acted as a hired gun to testify again and again that there was no problem with lead contamination.

As happened with the battles over smoking, ozone-destroying chlorofluorocarbons, and acid rain, and now fossil fuels that produce greenhouse gases, powerful industries will do all they can to discredit and even destroy scientists who are threatening their business with the truth. As the famous memo from the tobacco PR firm once said: "Doubt is our product." They employ lobbyists and hired-gun "experts" to sow doubt and uncertainty about the problem and manage to sway politicians with their testimony (plus their abundant campaign contributions).

Patterson endured repeated attacks and tremendous abuse from scientists who were paid by the oil and lead industries, and his research efforts were threatened. Luckily, he was tenured at Caltech and highly regarded among professional scientists not in the pay of corporations, so he never lost his lab or his job. However, many research organizations refused to fund his grants, including the U.S. Public Health Service (which is supposedly protecting public health, not corporate profits). As late as 1971, the National Research Council excluded him from their panel on atmospheric lead contamination, even though he was the world's leading expert on the topic.

Finally, however, the early 1970s marked a new awareness of environmental issues, and the tide began to turn. The Environmental Protection Agency was founded in 1972, and for the next few years (especially after Democrats took over Congress in 1974) almost any environmental law passed Congress with almost unanimous votes from both parties. Environmentalism was a popular bipartisan issue back then, and the Republican Party was not yet in the grips of powerful polluters. By 1975 the United

States had mandated that all cars use unleaded gasoline and catalytic converters, and by 1986 Patterson's work meant that all gasoline was lead-free. Meanwhile, Patterson showed the problem with contamination of lead in food, comparing the lead in canned fish with the lead levels in 1,600-year-old Peruvian skeletons of people who ate a lot of fish but were never exposed to lead in the environment.

By 1978 he was on the National Research Council panel, advocating for further reductions in lead in the environment—but the majority didn't agree with him, so he wrote a strongly worded 78-page minority report. But Patterson's heroic struggle for science and against polluters has done its work. By the late 1990s, the lead levels in the blood of average Americans had dropped by 80 percent. We can all thank this courageous scientist who started with a simple problem of dating the oldest meteorites in the solar system and ended up saving the planet. Patterson is an object lesson on scientific integrity and following the data wherever it may lead, no matter what powerful interests and hired-gun scientists oppose you, to promote the general good of the population and the planet.

FOR FURTHER READING

Bevan, Alex, and John De Laeter. *Meteorites: A Journey Through Space and Time.* Washington, D.C.: Smithsonian Books, 2002.

Chambers, John, and Jacqueline Mitton. *From Dust to Life: The Origin and Evolution of Our Solar System.* Princeton, N.J.: Princeton University Press, 2013.

Dalrymple, G. Brent. *Ancient Earth, Ancient Skies: The Age of the Earth and Its Cosmic Surroundings.* Stanford, Calif.: Stanford University Press, 2004.

Gargaud, Muriel, Hervé Martin, Purificacíon López-García, Thierry Montmerle, and Robert Pascal. *Young Sun, Early Earth, and the Origins of Life: Lessons for Astrobiology.* Berlin: Springer, 2013.

Nield, Ted. *The Falling Sky: The Science and History of Meteorites and Why We Should Learn to Love Them.* New York: Lyons, 2011.

Norton, O. Richard. *Rocks from Space: Meteorites and Meteorite Hunters.* Missoula, Mont.: Mountain Press, 1998.

Smith, Caroline, Sara Russell, and Gretchen Benedix. *Meteorites.* London: Firefly, 2010.

Zanda, Brigitte, and Monica Rotaru, eds. *Meteorites: Their Impact on Science and History.* Cambridge: Cambridge University Press, 2001.

11 GREEN CHEESE OR ANORTHOSITE? THE ORIGIN OF THE MOON

MOON ROCKS

That's one small step for a man, one giant leap for mankind.

—NEIL ARMSTRONG

ONE GIANT LEAP....

Like many Americans older than 55, I was glued to the TV set on July 20, 1969. I was visiting my cousins' ranch outside Hot Springs, South Dakota, at the time, getting a month's experience in ranch life: collecting eggs from the chickens, riding horses and tractors, taking care of the daily chores, and getting to know my extended family. We'd been hearing the buildup to the Apollo 11 mission all week, but now we were about to witness something extraordinary: the first man to walk on the moon, and even more amazing for that time, the world would be able to watch it live on television! We all clustered around the TV in the small living room that afternoon as the networks began to broadcast the preparations for the first moonwalk. Then, finally, the magic moment arrived, and millions of people all over the world simultaneously saw one of the most stirring achievements in human history.

The race to the moon was launched by President John F. Kennedy in 1961, challenging the United States, and especially its space program, to land a man on the moon before the end of the decade. We had been trailing the Soviets badly in the space race ever since they first launched the Sputnik satellite in 1957, long before we could do so. Then they launched the first animals into space, and then the first man into space, Yuri Gagarin, in 1961 (a month before American astronaut Alan Shepard). Between 1959 and 1963, the Mercury program launched the first Americans into space, and we were all riveted when John Glenn was the first American to orbit the earth in 1962. From 1965 to 1966, we moved on to the Gemini program, with two astronauts doing even more daring missions, including space walks and docking between spacecraft. From 1968 to 1972, the Apollo program with its three-man crews was building up the expertise to land on the moon and then return, each mission flying longer times and farther distances around the moon than before.

Finally, on that fateful day in 1969, Apollo 11 reached the moon. While the third astronaut, Michael Collins, remained in orbit around the moon, Neil Armstrong and Buzz Aldrin flew the lunar lander to the moon's surface, then did a short moonwalk (figure 11.1) before blasting off and returning to the mother ship for their voyage back to Earth. Each successive Apollo mission (Apollo 12 through Apollo 17, except for the ill-fated Apollo 13, which had an explosion in space and barely returned its astronauts alive) took longer and longer moonwalks and brought back more and more samples. By the time Congress canceled the Apollo program in 1973, the six moon missions had landed 12 men on the moon, collected a huge amount of data about the moon, and returned with 381.7 kilograms (842 pounds) of lunar samples. The only scientist ever to walk on the moon was a geologist, Harrison Schmitt, who was on the last mission, Apollo 17, which spent several days on the moon in December 1972.

The space program generated a huge program of research that produced enormous technological breakthroughs not only in space, but in all sorts of other fields as well. It jump-started the race for smaller and faster computers and hugely improved telephone communication, especially satellites for communication and GPS navigation. The robots assembled to build spacecraft eventually made our assembly lines for cars and many other products more efficient. A wide variety of products were developed based on NASA research, including artificial hearts, thermal blankets, strong but light metal alloys and lightweight composite materials, better drugs grown in zero gravity, smoke detectors, air-purification systems, small practical lasers, high-capacity batteries, UV sunglasses, Teflon-coated fiberglass, better fire protection gear for firefighters, solar power systems, artificial limbs, MRI and CAT scanning, LED technology, joysticks for video games, better golf balls, the TACS system that aircraft use to avoid collisions,

Figure 11.1 ▲
Photograph of Apollo astronaut Edwin Eugene "Buzz" Aldrin on the moon, taken by Neil Armstrong, July 20, 1969. (Courtesy of NASA)

virtual reality simulators, hydroponics, DirecTV, pacemakers, and even disposable diapers.

In another sense, the space program was immensely important not only in giving us satellite images of Earth to study all sorts of processes happening on our planet, but in providing the humbling views of Earth from space that transformed the way we think about our "pale blue dot." All this and more was produced for less than 1 percent of the federal budget, a trivial amount compared with what we spend on other things that provide much fewer benefits.

SISTER, DAUGHTER, OR PICKUP?

Perhaps the biggest scientific benefit of all was the definitive answer to the long-standing scientific question: How did the moon form? What was it made of? Various legitimately scientific ideas beyond the "green cheese" hypothesis had been floating around in the planetary geology and astronomy community for more than a century. The ideas fall into three broad categories, which (thanks to the male-dominated community of astronomers) acquired sexist names that are no longer acceptable:

1. *The "pickup" or "capture" hypothesis*: For decades, some scientists had suggested that the moon was a foreign body from far outside the earth's orbit that was captured as it flew by and pulled into orbit by the earth's gravity. But there were numerous problems with this model from the very start. For one thing, the moon's orbit moves in the same plane as the earth's orbit around the sun, which would be unlikely if an object coming at any angle from outer space were captured. Such an orbit would swing around the earth in any plane except the plane of the earth-sun system. In addition, the usual consequence of gravitational capture of a large body is either collision, or else the object flies back off into space with an altered orbit. For the moon to have been slowly captured by the earth's gravity, staying in orbit without collision or escape, the earth would have needed to have a very thick atmosphere that extended much farther than it does now. No evidence supports this. Finally, if the moon were an exotic object captured by Earth's gravity, its composition would be radically different than that of the earth. The moon rocks could be used to test this idea.
2. *The "daughter" or "fission" hypothesis*: This scenario, first proposed by astronomer George Darwin (son of Charles Darwin) in the late 1800s, argued that the moon is made of matter from the original rapidly spinning earth. During this rapid spin, molten material spun off from the earth into space to form the moon. Some astronomers even suggested that the Pacific Ocean Basin was the remnant scar of that event. In 1925 Austrian geologist Otto Ampherer proposed that the spinning off of the moon caused continental drift. This scenario was plausible for many years, although by the 1960s plate tectonics had shown that the Pacific Basin is not an ancient scar but floored by very young lavas less than 160 million years old. This model also didn't account for the angular momentum of the earth-moon system.

Once again, the crucial test would be the moon rocks. If they were the same composition as the primordial earth (before it separated into core, mantle, and crust), then this concept would be plausible.

3. *The "sister" hypothesis*: Similar to the "daughter" hypothesis, this model suggests that the original earth-moon system started out as two large blobs of matter that became locked into gravitational attraction with each other. Again, there are problems with the angular momentum of the earth-moon system in this model. But like the "daughter" hypothesis, it predicts that moon rocks would have a composition very similar to the primordial earth.

These ideas and more were hanging in the balance when Apollo 11 and later moon missions brought lunar samples back to labs to study. To everyone's surprise, their composition did not support any of the previous ideas. Instead, it suggested a new idea that no one had ever imagined.

IMPACT!

The lunar rocks brought back by the Apollo missions (figure 11.2) were not similar to the early earth in composition. Nor were they some exotic composition, as if they had been a body from outside the earth captured by gravity. Instead, they were made of anorthosite and its volcanic equivalent, the familiar black lava known as basalt. In other words, their composition was very much like parts of the upper mantle where the lavas that erupt onto the seafloor or from volcanoes like Kilauea on Hawaii have their source.

This was a shock. If the moon was made almost entirely of mantle material, it must be a piece of the earth's mantle that formed after the primordial earth had separated into a core of iron and nickel (chapter 10) and a mantle made of silicate minerals. In other words, the moon was formed long after the earth had cooled and coalesced and its layers had differentiated and separated.

Even more startling, the only way to get a lot of mantle material into space was to blast the early earth with a giant impact from another body (figure 11.3). Planetary geologists now call this body Theia (the Greek name for the mother of Selene, the moon goddess) and postulate that it was a Mars-sized protoplanet that hit the earth with an impact that blew material sideways off the earth and into orbit. Once this debris began to orbit the earth (at one-tenth the distance from the earth that the moon is today),

Figure 11.2 ▲
Sample of typical lunar anorthosite. (Courtesy of Wikimedia Commons)

Figure 11.3 ▲
An artist's conception of the impact that blasted the mantle and formed the moon. (Courtesy of Wikimedia Commons)

it would have gradually coalesced. The energy of this collision would have been amazing! Trillions of tons of material would have been vaporized, and the temperature of the earth would have risen to 10,000°C (18,000°F).

The heat from its own radioactive minerals would have melted the moon completely, and most of the moon would have remained the same composition as the earth's mantle, while the melting would also have caused huge eruptions of basaltic lava, forming magma oceans that now make the dark "maria," or "seas," on the moon's surface (figure 11.4). Meanwhile, the moon has a tiny iron core, only 330–350 kilometers in diameter, thought to be a relic of the core of Theia left behind after the collision; most of its

Figure 11.4 ▲

The near side of the moon, which always faces Earth, showing its impact craters and dark lava flows ("maria"). (Courtesy of NASA)

iron-nickel core accreted to the earth's core. By contrast, if the "sister" or "daughter" models (favored before Apollo 11) were correct, the moon would have a large core, roughly proportional to the size of the earth's core relative to its mantle.

When did this all occur? Once again, moon rocks give the answer. Using the same uranium-lead and lead-lead dating methods discussed in chapter 11, many labs have dated moon rocks. Most are at least 4 billion years old, suggesting that the moon's surface formed early and has not been modified much since. After all, it has none of the forces that change the earth's surface—it has no atmosphere, no water, no weathering, very reduced gravity, and no plate tectonics. The only major modification of its surface has been huge impacts that left craters (figure 11.4), and most of the crater debris has been dated at older than 3.9 billion years, so most of the impacts occurred early, and not much has happened since then.

The oldest pre-impact rock dates from the moon are currently 4.527 ± 0.0010 billion years. This is about 30 million years younger than the meteorites that date back to the origin of the solar system, so the moon is definitely younger than the events that formed the solar system and the earth and the melting and differentiation episode that separated the earth's core from its mantle.

Since the initial proposal of the giant impact hypothesis, much additional evidence has come out of analysis of moon rocks to support the mantle source of the moon. Nearly all the geochemical isotopes (oxygen, titanium, zinc, and many others) that have been studied in the past 48 years since the moon rocks were collected have shown that the moon and the earth's mantle have identical chemical compositions. There are also many refinements to the impact model, with some versions having more than one impacting body or positing different-sized impactors or different impact mechanics. But no matter which version is currently favored by scientists, the Apollo samples inescapably point to the moon as being a chunk of the earth's mantle.

MOONSTRUCK!

It's funny. When we were alive we spent much of our time staring up at the cosmos and wondering what was out there. We were obsessed with the moon and whether we could one day visit it. The day we finally walked on it was

> celebrated worldwide as perhaps man's greatest achievement. But it was while we were there, gathering rocks from the moon's desolate landscape, that we looked up and caught a glimpse of just how incredible our own planet was. Its singular astonishing beauty. We called her Mother Earth. Because she gave birth to us, and then we sucked her dry.
>
> —Jon Stewart, *Earth (The Book): A Visitor's Guide to the Human Race*

People have stared up at the moon for centuries and imputed mysterious powers to it. In the hit movie *Moonstruck*, the characters behave strangely under the influence of the full moon. Supposedly full moons cause werewolves to change out of their human form or real humans to act crazy. In astrology, the moon's position at the moment we are born supposedly affects our personality and future, although this is pure bunk. Many cultures blamed the moon for odd events or worshiped it as a god. Early science fiction imagined men on the moon or aliens from the moon invading Earth. Lots of cultures have looked up at the blotchy pattern of dark and light surfaces and imagined a "face" or seen the "man in the moon." In one of the earliest (1902) silent movies ever made, *A Trip to the Moon*, the main characters are loaded into a cannon and shot to the moon, where the cannon shell sticks in the "eye" of the "man on the moon." It was influenced by Jules Verne's science fiction novel, *From the Earth to the Moon* (1865), which blasted the explorer to the moon with a cannon. None of these ideas makes sense any more with our current understanding of the moon.

However, there are real and surprising ways the moon does affect us. Astronomers and physicists point to an interesting dynamic: the relatively large size of our moon (compared to the satellites of other planets) acts as a stabilizer, keeping the earth's rotation fairly steady so it doesn't flop on its side, like Neptune has. Many also think that the reason the rotational axis is tilted 23.5° from the plane of the orbit around the sun is due to the impact as well. When the impact occurred, it knocked the earth's rotational axis 23.5° off vertical, so now it wobbles like a spinning top (see chapter 25).

Just as amazing as the original impact is the story of how the earth-moon system got to be the way it is now. Today, the tidal pull of the earth has completely stopped the moon's rotation on its axis, so it is tidally locked to always show the same face to the earth. This is the only side of the moon that any

Figure 11.5 ▲
The far side of the moon, only visible to spacecraft that orbit the moon, starting with Apollo 8 in 1968. (Courtesy of NASA)

human saw until the Apollo 8 mission first flew around the opposite side of the moon (figure 11.5) and photographed it. (This is not the same as Pink Floyd's "Dark Side of the Moon," which is a myth; there is no permanent "dark side," since both sides experience darkness and light depending upon the position of the sun.) Meanwhile, the moon's tidal pull on the earth has been gradually slowing the earth's rotation, making it lose about 1.5 milliseconds every century, and more than a minute in a few thousand years. At the beginning of each New Year (especially after the millennium in 2000), the world's most precise clocks need to be adjusted to account for this, since otherwise they would be out of synch with the atomic clocks.

A few milliseconds a year might not seem like much, but over millions and billions of years, it adds up. Physicists have done the calculations, and found that the earth slowed down so much that there were far more rotations of the earth (Earth days) in the geologic past than there are now.

Confirmation of this startling idea came from the paleontology of humble corals. In the early 1960s, paleontologist John W. Wells of Cornell University was looking at fossil corals that had both daily growth lines and larger marks that showed the annual cycle of the seasons. He was able to slice the corals very thin, polish them, and then count the growth rings under the microscope. Sure enough, in the Devonian Period (about 400 Ma), the earth spun on its axis so much faster that it had 400 turns (400 days) in one revolution around the sun (1 year). About 600 Ma, the day was only 21 hours long, not 24 hours, and there were 430 days in a year. And only 150 Ma, there were 380 days in a year.

What does this mean? Although it's extremely slow by human standards, the earth is gradually slowing down, so some day about 20 billion years from now, it will also be tidally locked, and only one side will face the moon and one side will never see the moon. The energy of the entire earth-moon system is also decreasing, so both systems are slowly pulling apart. When the moon debris first was blasted out into space, the moon was only 10 percent of its current distance away from the earth, and it has been slowly receding since then. Back when the trilobites roamed (600 Ma), the moon was much closer and would have looked huge in the sky. Its tidal pull at that distance would have been so powerful that immense true tidal waves (not tsunamis formed by earthquakes, which have nothing to do with tides) would have swept across the earth as the tides rose and fell.

Eventually, the two bodies will not only be tidally locked and much farther apart, but their motion could come to a halt. However, this is billions of years in the future, and the sun will probably explode before then and wipe out the inner planets, so it will never actually get to occur.

It's pretty humbling when you think about it. Just a few pounds of rocks brought back by the Apollo spacecraft have revolutionized our understanding of the moon, the earth, and the solar system. The next time you read poems about the moon, or hear romantic lyrics about the "moon in June," you'll never think about our only natural satellite the same way again.

FOR FURTHER READING

Chaikin, Andrew. *A Man on the Moon: The Voyages of the Apollo Astronauts*. New York: Penguin, 2007.

Chambers, John, and Jacqueline Mitton. *From Dust to Life: The Origin and Evolution of Our Solar System*. Princeton, N.J.: Princeton University Press, 2013.

Dalrymple, G. Brent. *Ancient Earth, Ancient Skies: The Age of the Earth and Its Cosmic Surroundings*. Stanford, Calif.: Stanford University Press, 2004.

French, B. M. *Origin of the Moon: NASA's New Data from Old Rocks*. Greenbelt, Md.: NASA Goddard Space Flight Center, 1972.

Gargaud, Muriel, Hervé Martin, Purificacíon López-García, Thierry Montmerle, and Robert Pascal. *Young Sun, Early Earth, and the Origins of Life: Lessons for Astrobiology*. Berlin: Springer, 2013.

Harland, David M. *Moon Manual*. London: Haynes, 2016.

Hartmann, William K. *Origin of the Moon*. Houston: Lunar & Planetary Institute, 1986.

Mutch, Thomas A. *Geology of the Moon: A Stratigraphic View*. Princeton, N.J.: Princeton University Press, 1973.

Reynolds, David West. *Apollo: The Epic Journey to the Moon, 1963–1972*. New York: Zenith, 2013.

12 | EARLY OCEANS AND LIFE? EVIDENCE IN A GRAIN OF SAND

ZIRCONS

To see a World in a Grain of Sand
And a Heaven in a Wild Flower,
Hold Infinity in the palm of your hand,
And Eternity in an hour.

—WILLIAM BLAKE, "AUGURIES OF INNOCENCE"

I do not know what I may appear to the world, but to myself I seem to have been only like a boy playing on the sea-shore, and diverting myself in now and then finding a smoother pebble or a prettier shell than ordinary, whilst the great ocean of truth lay all undiscovered before me.

—ISAAC NEWTON

MORE PRECIOUS THAN DIAMONDS

If you flip channels and watch some of those home-shopping shows or infomercials on TV, sooner or later you see someone hawking gaudy jewelry made of "cubic zirconia crystals." They ooh and ahh about the diamond-like appearance of the cubic zirconia crystals, for sale at a fraction of the cost of real diamonds—but who can tell? Likewise, you can find many vendors online touting their cubic zirconia jewelry with texts like this:

> Ziamond is where discerning customers shop for the best high quality cubic zirconia jewelry exclusively set in precious metals like 14k gold, 18k gold and luxurious Platinum. Our cubic zirconia jewelry and gems are the epitome of the finest lab created man-made diamond simulants that the world has to offer and are backed with our comprehensive Lifetime Guarantee.

> Ziamond's amazing cubic zirconia jewelry and cz stones are precisely cut then polished to fine diamond standards, ensuring our customers the true look and feel of a brilliant genuine diamond. You can wear Ziamond's cubic zirconia jewelry daily with confidence and clean it just as if you would your fine diamond jewelry. With over a century of fine jewelry experience our staff has the technical and artistic skills to create and execute any cubic zirconia jewelry design while maintaining attention to details, design quality and craftsmanship expected with fine jewelry. Whether it's diamond look cubic zirconia rings, cubic zirconia wedding rings, cubic zirconia engagement rings, cubic zirconia bracelets or cubic zirconia earrings, we look forward to providing you the Ziamond experience and see for yourself why we are recognized as a leader in our field as the best cubic zirconia jewelry company.

There's nothing wrong with cubic zirconia if you want a cheap faux diamond to impress your friends. But it's not a real diamond, and zirconium is not even that rare in nature. Cubic zirconia is just a synthetic gem made of zirconium dioxide (ZrO_2). (In its natural mineral form, ZrO_2 is known a baddeleyite, after Joseph Baddeley, a superintendent of a railroad project in Sri Lanka who first discovered it. Not a name you would hear the infomercial announcers using to sell jewelry.) Another mineral formed from the element zirconium is zirconium silicate ($ZrSiO_4$), know as the mineral zircon (figure 12.1). Large crystals of zircons form an octahedron, a structure that looks like two pyramids glued together. They come in many different colors, including purple, yellow, pink, red, and clear, depending upon what impurities are in them and what has happened to their crystal structure.

But although diamonds have great commercial value, the scientific information content of zircons is much greater, making them more precious scientifically than any diamond. Zircons are incredibly useful minerals to geologists. They form very hard, durable crystals, especially in granitic magmas during the last stage of cooling. Because zircon crystals have spaces for big atoms like zirconium, they also trap other large atoms that won't fit in any other mineral and that were concentrated in the last stage of magma crystallization. These include extremely rare elements like uranium and thorium. Thus, you can take a crystal of zircon and analyze it for its uranium content. Zircons are widely used to produce dates in the

Figure 12.1 ▲
A crystal of zircon. (Courtesy of Wikimedia Commons)

uranium-lead or lead-lead method of dating and in another method known as fission-track dating.

Zircons are so durable and resistant to nearly everything that brute force methods must be used to extract them from granitic rocks. Typically, you use a crusher to reduce the original rock to a fine powder. Then you soak the powder in the most powerful acid on Earth, hydrofluoric acid (HF). It is so caustic you must work in a very good fume hood and wear lots of protection for your skin, eyes, and lungs. HF even dissolves many types of containers, so you have to keep it in special bottles. HF will break down nearly every other mineral in the rock except the zircons, so once the acid bath is finished, you rinse the residue with water, and you have concentrated zircons, ready for analysis.

Zircons are durable not only in the lab setting but in nature as well. When rocks weather into sand grains that go bumping and bashing in the sand along the bottom of a stream, zircons are among the most durable minerals. Even in the most heavily weathered sand, which is about 99 percent quartz (the most common and durable mineral on the earth's surface), there will still be a small fraction of a percent in zircons as well. In fact, the presence of zircon (plus two other durable minerals in sand, tourmaline and rutile) has been used as a "ZTR index" to measure how extremely weathered and winnowed the sand or sandstone is. There are many geologists who specialize in zircons because they can be very powerful tools for all sorts of geological problems.

WHO'S ON FIRST?

Zircons are particularly useful, because geochronologists find that they are often the best mineral for dating very ancient rocks. The dates on some of the meteorites (chapter 10) and moon rocks (chapter 11) came from zircons, and zircons work for dating ancient Earth rocks as well. Many of the oldest rocks on Earth have been dated by applying not only uranium-lead methods but also lead-lead and rubidium-strontium dating to zircons.

For many years, the oldest known rocks on Earth were the Amitsôq Gneisses (figure 12.2) from the Isua Supracrustal belt on the southwest coast of Greenland, which gave dates of 3.8 billion years. They represented some of the earliest crustal rocks formed, including small blocks of proto-continental crust (now metamorphosed into gneisses), plus slices of ancient proto-oceanic crust (known as greenstones), and even some pieces of the earliest mantle (peridotites). However, this was not the maximum age, because these rocks had been highly metamorphosed and altered, so it was always possible that their true age was quite a bit older. What they did tell us was that the earliest crust of the earth was made of very small blocks of continental crust (proto-continents) that floated in a very thin, hot oceanic crust made of lavas that erupted directly from the mantle. These weird lavas, known as komatiites, were made entirely of mantle minerals like the greenish silicate called olivine. They are even richer in magnesium and iron than the basaltic lavas that make up all the oceanic crust today. They tell us

Figure 12.2 ▲

The Isua Supracrustals of southern Greenland, dated at 3.8 billion years old. (Courtesy of Wikimedia Commons)

that the early earth was still very hot and that the crust was thin and highly mobile and easily remelted. It still had not differentiated into the mature types of oceanic and continental crust rocks we have today. In fact, the crustal blocks were so small and thin that true plate tectonics probably did not exist yet. Komatiite lavas could only form under these conditions and no longer erupt anywhere on Earth now that oceanic crust has matured and the temperature and chemistry of the upper mantle has changed. Today, only lavas that cool to form basalts erupt to form the ocean floor.

Then, in 1999, another ancient rock, known as the Acasta Gneiss (figure 12.3), pushed the record for the earth's oldest rock back from 3.8 billion years to 4.031 ± 0.003 billion years. The Acasta Gneiss is another piece of proto-continental crust, but it is from a block known as the Slave Terrane, which got its name from Great Slave Lake in the Northwest Territories of Canada. This rock was mentioned in all the textbooks and held the record for a number of years.

Figure 12.3 ▲

The Acasta Gneiss near Great Slave Lake, Canada, dated at 4.01 billion years old. (Courtesy of Wikimedia Commons)

Just as in athletics, however, records are meant to be broken. In 2008 the Nuvvuagittuq Greenstone belt in northwest Quebec, on the east shore of Hudson Bay, gave dates of 4.28 and 4.321 billion years. This was determined not by direct uranium-lead dating, but by samarium-neodymium dating of the lavas in the greenstone belt. However, this date is controversial. Many scientists think that the dates of 4.28 billion years and older are not the age of the rocks but the age of the parent material that was remelted to form these rocks. The oldest uranium-lead date on the zircons from the rocks themselves suggest that they are really only around 3.78 billion years old. Even if this is true, it is evidence of the oldest crustal formation around 4.28 to 4.32 billion years ago. Given the history of research, we can expect some geologist to find a rock even older before long.

Notice that the oldest Earth rocks are no older than 4.32 billion years, yet the oldest materials of the solar system (meteorites and moon rocks) are at least 4.55 billion years old. Why the difference? The answer is plate tectonics, and the deep weathering of the earth's surface caused by water and wind. The earth's surface is constantly being recycled and remodeled by the motion of plates melting and plunging into the mantle, then being born again. The moon, by contrast, has a dead surface with no plate tectonics, so some of its rocks date to 4.5 billion years ago when it formed. Meteorites that formed with the early solar system, like the carbonaceous chondrites discussed in chapter 10, have been unaltered since they cooled, so they give the oldest dates of all.

COOL EARTH

These are the dates for the oldest rocks on Earth, but they are not the oldest Earth materials known. That distinction goes to a handful of zircon sand grains (figure 12.4) from a much younger sandstone found in the Jack Hills of Western Australia. Each individual grain can be dated by uranium-lead

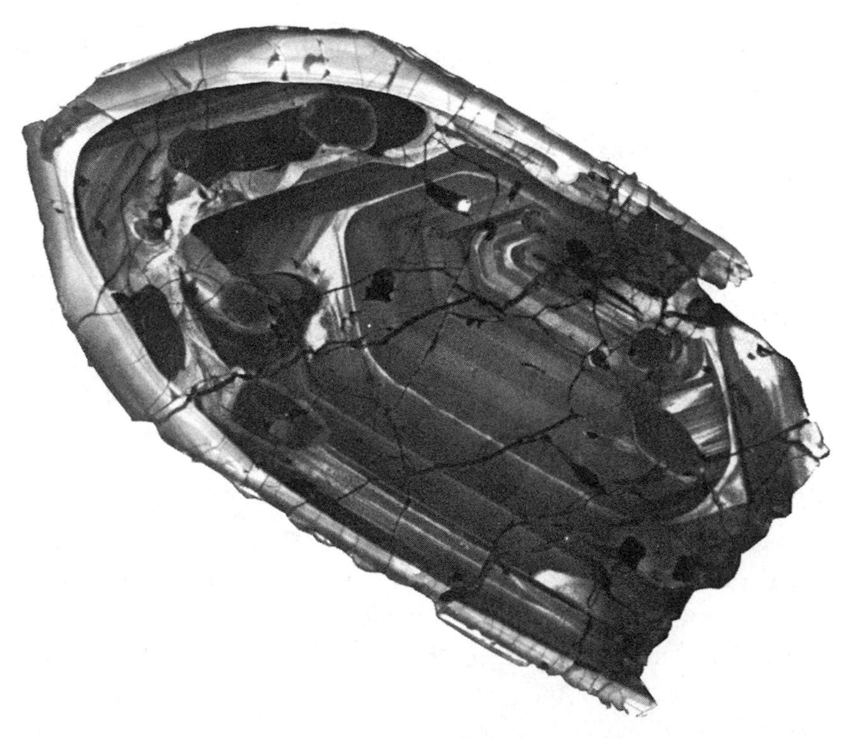

Figure 12.4 ▲

Microphotograph of a zircon grain from the Jack Hills of Australia, which gives a date of 4.4 Ga. It provides evidence that the earth was covered by liquid water at that early date. (Courtesy of J. Valley)

methods, so they give a scatter of ages. But the oldest grains of all give an age of 4.404 billion years, at least 100 million years older than the 4.3-billion-year-old materials from Quebec. Thus, the current record holder for the oldest material from Earth (that is, not a meteorite or moon rock) is 4.4 billion years. These sand grain dates put us closer and closer to the age of moon rocks and meteorites, but we still have a gap of about 200 million years. This is about the same time span as the beginning of the Age of Dinosaurs (Late Triassic) until today, so it is not a trivial amount of time.

But those same tiny zircon sand grains held even more surprises. Not only did they give the oldest known dates, but when scientists analyzed the tiny bubbles of gases trapped inside them, they found evidence of the early atmosphere from over 4 billion years ago. These bubbles had oxygen isotopes in them that suggested the earth had liquid water on its surface as early as 4.4 billion years ago!

Prior to this discovery geologists had always assumed that the earth took a long time to cool from its molten state at 4.55 billion years ago. Most thought that the earth took about 700 million years to cool down below the boiling point of water (100°C), because that was the age of the oldest sedimentary rocks that would have been produced by running water (the Isua Supracrustals from Greenland mentioned above, which are 3.8 billion years old). But the Jack Hills zircons turn that assumption inside out. If they truly indicate the presence of liquid water on the earth 4.4 billion years ago, then it took only 200 million years for the earth to cool from its molten state to a condition that was below the boiling point of water. This also suggests that there were not as many meteorite impacts during this time interval, or the oceans would have been vaporized over and over again. Taken together, these data suggest what is now called the "cool early Earth hypothesis."

So where did this early Earth water come from? Traditionally, geologists thought that it was water trapped inside the earth's mantle when it cooled, gradually escaping through volcanoes in a process called degassing. But lately, chemical analyses of extraterrestrial objects match the chemistry of the earth's oceans (especially carbonaceous chondrite meteorites—see chapter 10). This suggests that there was a lot of water trapped in the debris of the early solar system (of which the chondrites are remnants). The same is true of moon rocks, which do not have much water in them today, but apparently were pretty wet when the solar system formed. If this is so, then the earth was born with its water already present as it cooled and condensed. It only required its surface temperature to drop below 100°C for that water to form the first oceans.

One thing we can rule out is comets. Although comets are often called "dirty snowballs" because they are made mostly of dust and water ice, chemical analyses of four comets now show that their geochemistry is very different from the earth's water. Thus, the popular idea that comets impacted the early earth and melted to form its oceans can be dismissed.

Those tiny zircon sand grains from the Jack Hills had one more surprise in them. In 2015 a paper was published on the tiny crystals of graphite that were also trapped inside them. Graphite is more familiar to most people as a mineral form of crystalline carbon, the same mineral that makes your pencil "lead." Amazingly, the geochemical data from that graphite was consistent with the isotope ratios of carbon found in life! These particular zircon grains gave dates of 4.1 billion years, so they were not as old as the oldest zircons with water chemistry in them at 4.4 billion years. Nevertheless, this is a startling piece of data. Previously, the oldest carbon that had the right chemistry to be produced by life, as well as possibly the oldest fossils, came from those Isua rocks from Greenland, dated at 3.8 billion years. The Jack Hills zircons are 300 million years older than the previous record holder. And the oldest strong fossil evidence of life comes from the Apex Chert in the Warrawoona Group of Western Australia, dated at 3.5 billion years, and the Fig Tree Group of South Africa, dated at 3.4 billion years. So this extends the origin of life much earlier than previously supposed and not that much later than the early oceans on the cool early earth.

Once more, this much older evidence of life, just like the evidence of early oceans from the same zircons, forces us to revise our ideas about the early earth. Based on the dates of the moon craters (which all cluster between 3.9 and 4.4 billion years in age), we assumed that the early earth must have also undergone intense bombardment of leftover debris from the solar system prior to 3.9 billion years ago. But the evidence of liquid water oceans at 4.4 billion, and possibly even life at 4.1 billion, makes it appear that the bombardment of Earth was much less intense than previously supposed.

By the time this book is published, the odds are good that even more amazing discoveries will be announced, and there may be a yet older candidate for the oldest rock on Earth. But that is a good thing. That is a sign that the science of early earth geology is an active and vibrant field, always yielding groundbreaking discoveries. For some, it may be frustrating to write a book that is out of date before it is published. But science marches on, and we are always learning new and more surprising things about the earth every time another critical analysis is published.

FOR FURTHER READING

Chambers, John, and Jacqueline Mitton. *From Dust to Life: The Origin and Evolution of Our Solar System*. Princeton, N.J.: Princeton University Press, 2013.

Gargaud, Muriel, Hervé Martin, Purificacíon López-García, Thierry Montmerle, and Robert Pascal. *Young Sun, Early Earth, and the Origins of Life: Lessons for Astrobiology*. Berlin: Springer, 2013.

Hazen, Robert M. *The Story of the Earth: The First 4.5 Billion Years from Stardust to Living Planet*. New York: Penguin, 2013.

Shaw, George H. *Earth's Early Atmosphere and Oceans, and the Origin of Life*. Berlin: Springer, 2015.

Ward, Peter, and Joe Kirschvink. *A New History of Life: The Radical New Discoveries About the Origin and Evolution of Life on Earth*. New York: Bloomsbury, 2015.

13 | MICROBIAL CONDOS: CYANOBACTERIA AND THE OLDEST LIFE

STROMATOLITES

If the theory [of evolution] be true, it is indisputable that before the lowest Cambrian stratum was deposited, long periods elapsed . . . and the world swarmed with living creatures. [Yet] to the question why we do not find rich fossiliferous deposits belonging to these earliest periods . . . I can give no satisfactory answer.

—CHARLES DARWIN, *ON THE ORIGIN OF SPECIES*

DARWIN'S DILEMMA

When Darwin published *On the Origin of Species* in 1859, one of the weakest lines of evidence was the lack of undisputed fossils before the Cambrian Period, when complex multicellular animals like trilobites first appeared. Darwin had personal familiarity with Cambrian rocks and fossils, because in 1831 he had been a field assistant in the Cambrian rocks of western Wales for his Cambridge mentor, the legendary geologist Adam Sedgwick, the first man to hold the title "professor of geology." But even when Darwin wrote about the topic in 1859, some 30 years later, the lack of fossils was still a mystery.

Darwin and most geologists since have known that one of the problems was that old rocks from before the Cambrian were usually transformed by enormous heat and pressure into metamorphic rocks, so all fossils were destroyed. In addition, the older the rocks are, the more likely they have not only been metamorphosed, but simply eroded away. Finally, really ancient rocks are usually at the bottom of the stack, and thus are buried by younger post-Cambrian sediments, so they are only exposed in a few places on the earth where the ancient "basement" rocks have been uplifted and erosion has stripped away their cover.

Nevertheless, scientists took up Darwin's challenge and kept on looking. There were many blind alleys and false leads. A weird branching structure that looked like a primitive plant was named *Oldhamia*. It turned out to be formed by the burrows of a worm, but it was not a body fossil. A slimy "organism" found in jars of mud from the deepest ocean and touted as a new creature, *Bathybius*, by Darwin's "bulldog" Thomas Henry Huxley turned out to be a product of chemical reactions of calcium sulfate with the alcohol used to preserve the sample. In 1858, the year before Darwin's book was published, pioneering Canadian geologist Sir William E. Logan discovered some layered structures on the banks of the Ottawa River just outside Montreal (figure 13.1). Most scientists were not convinced, because there are many nonbiological ways that layered structures can form in rocks. One of Logan's protégés, Canadian paleontologist J. W. Dawson, was convinced it was produced by life, and named the structures *Eozoon canadense* ("dawn animal of Canada"). He called it "one of the brightest gems in the crown of the Geological Survey of Canada." But soon other geologists looked closer at the "fossil" and where it had come from and concluded that it was a banded metamorphic structure formed by layers of calcite and serpentine minerals.

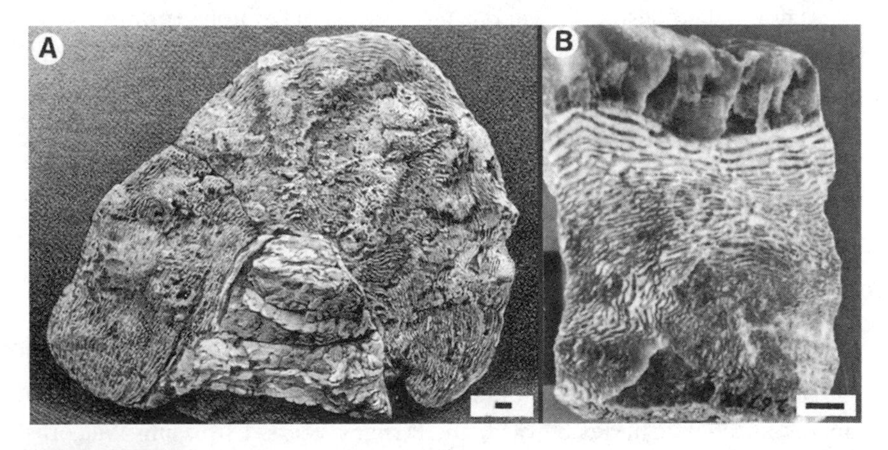

Figure 13.1 ▲

Layered structure known as *Eozoon* once thought to be a fossil, but now known to be inorganically grown pseudofossils. (*A*) Illustration in J. W. Dawson's book, *Dawn of Life*. (*B*) The holotype specimen in the Smithsonian Institution. Scale bar = 1 cm. (Photo courtesy of J. W. Schopf)

PSEUDOFOSSIL OR REAL FOSSIL?

After so many false alarms over pseudofossils, geologists were justifiably skeptical of any specimen touted as evidence of Precambrian life. It's an easy mistake to make. There are lots of ways that natural rocks and minerals can form structures that look (to the inexperienced collector) like real fossils. Many rock hounds split open shales and find delicate branching lacy black structures on them and think they have a fossil plant. But this is a well-known pseudofossil, known as a pyrolusite dendrite, produced by the branching growth of crystals of manganese oxide. Many a paleontologist has been handed an odd-shaped rock by an amateur and been told it's a "fossil egg" or "fossil brain" or "fossil heart" or even a "fossil phallus." Most of the time, these are simply sediments that have been cemented together into suggestive shapes, known as concretions.

But in 1878, another interesting structure was proposed as evidence of Precambrian life. The young Charles Doolittle Walcott was working as an assistant for the legendary James Hall, the first official geologist and paleontologist for the state of New York. (Walcott later became America's foremost paleontologist, and at various times he also served as the head of the Smithsonian Institution, director of the U.S. Geological Survey, and president of the National Academy of Sciences.) While visiting the resort and horse-racing mecca of Saratoga in the upper Hudson Valley, Walcott stopped at a place now called Lester Park, about 3 miles west of Saratoga Springs. There he found a large outcrop of layered structures that looked like the heads of a bunch of cabbages that had been sliced through (figure 13.2). These structures had been discovered by geologists before and named "stromatolites" (Greek for "layered rock"). However, the specimens in Lester Park were extraordinary. The 28-year-old Walcott sat down to write and publish his first scientific paper about them, giving them the name *Cryptozoon* (Greek for "hidden life") and touting their biological origin.

Naturally, he met with a cool reception from the scientific community. They had been burned before with the layered structure Dawson called *Eozoon*, and now they were gun-shy. The world's most famous paleobotanist, Sir Albert Charles Seward, spent many years dismissing Walcott's *Cryptozoon*, and he had a lot of influence. He rightly argued that they had no organic structures or detailed plant tissues or anything that would rule out simple layers caused by mineral growth.

Figure 13.2 ▲

The layered cabbage-like stromatolites known as *Cryptozoon* from Lester Park, New York, with their tops sliced off by a glacier. (Photo by the author)

Nevertheless, more and more different kinds of stromatolites were being found and described. Some were not shaped like simple sliced cabbages, but were tall domed pillars. Still others had shapes such as tall cones (*Conophyton*) or convex layers flattened in the center (*Collenia*). Well-preserved stromatolites of many shapes were common in the late Precambrian rocks of Siberia, so Soviet geologists set about naming and describing many such structures. Still, the proof that they were truly organic and not some sort of geological structure was missing.

SHARK BAY

The only convincing way to show that stromatolites were really fossils was to find them living and growing in the present day. Yet nearly all known stromatolites were from the Precambrian, more than 550 million years ago, with a few minor exceptions. They had once been the most common visible fossil on the planet but had mysteriously vanished during the Cambrian, when the evolution of multicellular animals took off.

The breakthrough occurred during routine geological exploration of a relatively unknown region. Geologist Brian Logan of the University of Western Australia and other geologists were mapping the northern coast of Western Australia in 1956. They traveled up the coast and reached a salty lagoon called Shark Bay, about 800 kilometers (500 miles) north of Perth. As they explored the bay at low tide, they found a shallow area called Hamelin Pool that was covered by pillars almost a meter or more tall with domed tops (figure 13.3). They were a dead ringer for *Cryptozoon* and other structures that had been called stromatolites.

Once they looked more closely and took samples, they found that the Shark Bay structures were indeed made of millimeter-scale layers of fine sediment, just as were found in fossil stromatolites. And now they could discover what had made these mysterious layered structures. The top surface of each pillar was covered by a sticky mat of blue-green bacteria, or cyanobacteria, growing in the sun. Older books called them "blue-green algae," but they are not algae, which are true plants with eukaryotic cells with a nucleus and organelles. Cyanobacteria are prokaryotes without a discrete nucleus but with the internal chemistry needed for photosynthesis.

Once these structures were analyzed further, it was apparent how they got their layered structure. The mat of cyanobacterial filaments is sticky,

Figure 13.3 ▲
Modern domed stromatolites growing in Shark Bay, Australia. (Courtesy of Wikimedia Commons)

so as fine sediment washes over it and settles, it is bound into a layer of filaments. Then the filaments grow up through the coating of sediment to seek the light again, making a new sticky mat that accumulates even more sediment. This goes on every day, so if the conditions are right, you could have a structure with hundreds of daily layers. When the cyanobacteria die, they leave behind a structure of layered sediment, without the organic material or more plant-like structures that the paleobotanists had long been demanding.

PLANET OF THE SCUM

So why did stromatolites, which had dominated the earth for 3 billion years, seem to vanish about 500 Ma? It turned out that Shark Bay is exceptional in more ways than just having a colony of living stromatolites. It has a very narrow mouth with a sandbar blocking it, which restricts the flow of water in and out of the bay during the tidal cycle. In addition, it is located in a tropical region, so the rate of evaporation is very high. This means that the water in the bay is very salty (7 percent salt, twice the salinity of the ocean). This is too salty for most of the snails and other organisms that would otherwise eat up the films of algae and cyanobacteria that naturally grow on intertidal rocks around the world.

Once the Shark Bay discovery was published in 1961, the tide of opinion turned quickly, and soon most paleobotanists and geologists agreed that stromatolites were indeed true fossil structures. Over the years, they have been found growing in a few more places, and these localities all have one thing in common: the water is inhospitable to any other organisms, especially snails and others that might eat the sticky mats. I've walked on them growing in salty lagoons along the Pacific Coast of Baja California. They grow in the salty waters of the west coast of the Persian Gulf, and huge dome-topped pillars like those at Shark Bay also grow in the salty lagoons of Lagoa Salgada ("salty lagoon" in Portuguese) in Brazil. Among the few that grow in normal marine salinity are those found in Exuma Cays in the Bahamas, where the water currents are too strong for even marine snails to hang on.

So why were they the most common visible fossil from Precambrian rocks? Remember, when cyanobacteria first evolved over 3.5 Ga, they were the only form of life on earth. We have their fossils from a locality dated to 3.5 Ga in the Warrawoona Group in Western Australia, and other kinds of bacterial fossils also dated to 3.5 Ga in the Fig Tree Group in South Africa, so they got an early start. Just as this book was finished, there were even reports of possible stromatolites from the Isua Supracrustals in Greenland, which have been dated to 3.8 Ga (see chapter 12). But for more than 3 billion years, the fossil record shows that nothing bigger than single-celled microbes evolved, so these mats of cyanobacteria had no grazers to crop them through 80 percent of life's history. They ruled the planet, and as my friend Professor J. W. Schopf of the University of California–Los Angeles puts it, Earth was the "Planet of the Scum" (figure 13.4).

Not until the Early Cambrian did snails and other organisms evolve that could crop these mats of cyanobacteria and algae that had blanketed the shallow seafloor uncropped for 3 billion years. Once they began to do so, the stromatolites nearly vanished. Meanwhile, the seafloor was no longer covered by a mat of sticky algae and cyanobacteria, so other animals could burrow into the sediment for the first time, and this opened up a whole new range of niches for life to exploit.

Figure 13.4 ▲

The world was dominated by simple layered stromatolites for most of its history. (Drawing by Carl Buell)

This "lack of cropping" explanation is supported not only by the places where they live today (with no snails or other croppers) but also by where they occasionally reappeared in the geologic past. Microbial mats are always ready to spring back and flourish anytime their croppers are suppressed. After three of the earth's great mass extinctions (the end-Ordovician, the Late Devonian, and the biggest mass extinction of all at the end of the Permian Period), stromatolites returned in abundance during the post–mass extinction "aftermath" times when there were few survivors of the animals that were clobbered by mass extinction. In each case, stromatolites grew like weeds, taking advantage of the wide-open landscape with few opportunistic survivor species and flourishing whenever the creatures that ate them had been wiped out.

Earth was the Planet of the Scum for over 80 percent of life's history. There were no other visible life forms that could leave fossils, only microbial mats building condos of stone. All the evolution was taking place in microbes, which only leave fossils in the best of circumstances. What held back the development of multicellular life? The very fact that the seafloor was blanketed by microbial mats might have also been a barrier, but once grazing snails evolved, many other animals could find niches as well. A seafloor without a film of slimy bacteria can be exploited and burrowed into by trilobites and many deeper diggers, and we see just that evidence in the burrows of the Early Cambrian.

There are lots of other ideas out there, but most geologists agree that the low levels of atmospheric oxygen prevented multicellular organisms from getting very large. Ironically, the low level of oxygen (discussed in chapter 14) was overcome by the photosynthetic activities of those same cyanobacteria that made stromatolites. It took them almost 3 billion years to do it, but bit by bit they pumped out so much oxygen that it finally overwhelmed all the crustal rocks that absorb it and eventually this gas became abundant in the oceans and atmosphere. Once it did so, it killed off most of the anaerobic bacteria that can only live in low-oxygen settings, triggering an "oxygen holocaust." Eventually, oxygen levels got high enough that oxygen-breathing multicellular animals, like worms and trilobites, could evolve. In fact, most of the oxygen you are breathing now comes not from trees in the forest but from the huge bloom of photosynthetic algae and bacteria in the oceans. So the next time you see some algal scum on a beach rock, thank it. You would not be here, nor could you breathe, without that scum.

FOR FURTHER READING

Chambers, John, and Jacqueline Mitton. *From Dust to Life: The Origin and Evolution of Our Solar System*. Princeton, N.J.: Princeton University Press, 2013.

Gargaud, Muriel, Hervé Martin, Purificacíon López-García, Thierry Montmerle, and Robert Pascal. *Young Sun, Early Earth, and the Origins of Life: Lessons for Astrobiology*. Berlin: Springer, 2013.

Hazen, Robert M. *The Story of the Earth: The First 4.5 Billion Years from Stardust to Living Planet*. New York: Penguin, 2013.

Knoll, Andrew H. *Life on a Young Planet: The First Three Billion Years of Evolution on Earth*. Princeton, N.J.: Princeton University Press, 2003.

Schopf, J. William. *Cradle of Life: The Discovery of Earth's Earliest Fossils*. Princeton, N.J.: Princeton University Press, 1999.

Shaw, George H. *Earth's Early Atmosphere and Oceans, and the Origin of Life*. Berlin: Springer, 2015.

Ward, Peter, and Joe Kirschvink. *A New History of Life: The Radical New Discoveries About the Origin and Evolution of Life on Earth*. New York: Bloomsbury, 2015.

14 | MOUNTAINS OF IRON: THE EARTH'S EARLY ATMOSPHERE

BANDED IRON FORMATION

With the mining sites, I found a subject matter that carried forth my fascination with the undoing of the landscape, in terms of both its formal beauty and its environmental politics.

—DAVID MAISEL

THE IRON RANGES

A trip to the Iron Ranges (figure 14.1) of northern Minnesota, Michigan, or southern Ontario, is a real eye-opener. If you visit the Mesabi Iron Range, or other major ranges such as the Vermilion and Cuyuna Ranges in Minnesota, or Gunflint Range stretching into Canada, or the Marquette and Gogebic and other ranges in the Upper Peninsula of Michigan, you see amazing sights. Entire mountains made mostly of iron have been removed, leaving immense open-pit mines that are now flooded with water. The Hull-Rust-Mahoning iron mine, near Hibbing, Minnesota, is one of the largest mines in the world (figure 14.2). Its pit alone spans 2.4 kilometers by 5.6 kilometers, and it is more than 180 meters deep (1.5 miles by 3.5 miles, more than 600 feet deep). When you stand on the rim, the pit below looks like a small ocean basin. As you look across the water flooding the bottom, you might see the modern operations still working. The bucket of the dragline to remove overburden is bigger than a house, and the mining equipment is on a heroic scale. The gigantic excavators and dump trucks (figure 14.3) have tires that are more than 3.5 meters (12 feet) in diameter.

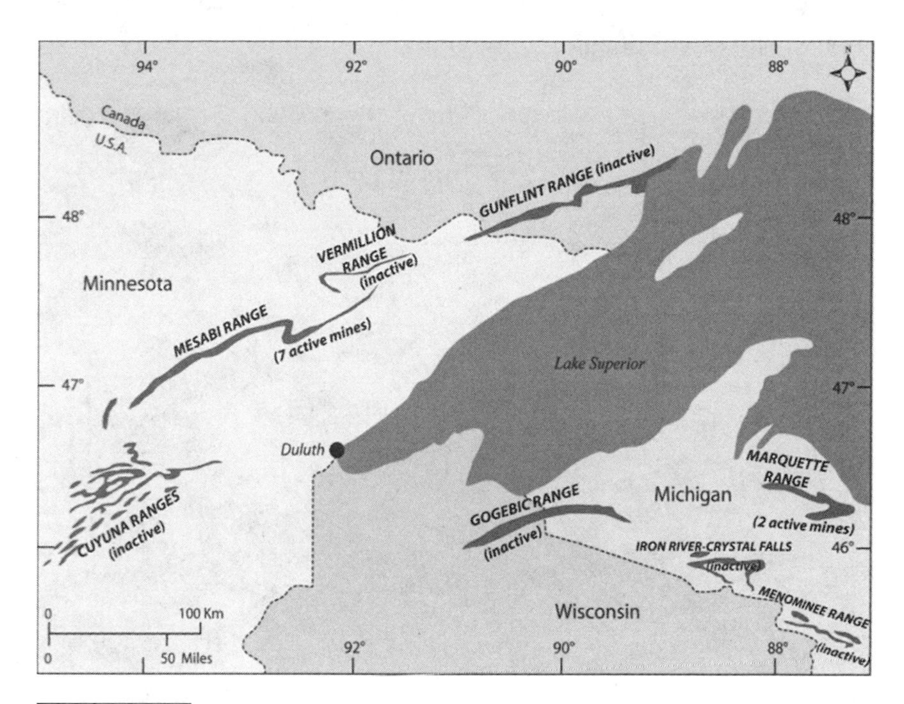

Figure 14.1 ▲

Map of the location of the iron ranges near Lake Superior. (Courtesy of Wikimedia Commons)

Figure 14.2 ▲

Panorama of the huge open pit of the Hull-Rust-Mahoning Mine, Minnesota. (Courtesy of Wikimedia Commons)

Figure 14.3 ▲
The size of the machinery in these huge mines is staggering. This is a retired dump truck from the Minnesota Mining Museum in Chisholm, Minnesota. (Courtesy of Wikimedia Commons)

Opened in 1895, the Hull-Rust-Mahoning iron mine has produced over 635 million metric tonnes of ore, and more than 450 metric tonnes of waste rock lies piled up all over the barren, desolate landscape. The town of Hibbing even had to be moved from its original site as the mine expanded and swallowed the original location of the town. Almost 25 percent of all the iron produced in the United States came from this hole in the ground that used to be a mountain of iron. Much of the iron that built the structures and machines of the Industrial Revolution in the late 1800s and early 1900s came from this mine, especially during the huge demand for steel for ships, tanks, and aircraft during the First and Second World Wars.

The Hull-Rust-Mahoning iron mine is not the only one of its kind. The Rochleau mine near Virginia, Minnesota, is 4.8 kilometers (3 miles) long, 0.8 kilometers (0.5 miles) wide, and 137 meters (450 feet) deep. Opened in 1893, it has produced more than 270 million metric tonnes of iron ore. It is still expanding, forcing the Mineview in the Sky mining museum to move to a new location. U.S. Route 53 had to be rerouted away from the expanding pit, with a bridge spanning part of the chasm. Other huge mines are found near Soudan, Minnesota. This town was waggishly named by miners who were daydreaming of the hot desert temperatures of the Sudan in Africa while they suffered through long, bitter, freezing Minnesota winters.

These mines and their wealth have had a huge influence on American history. The iron of the Lake Superior region meant that American steel could be used to build our great buildings and ships, along with millions of automobiles and other machines. Iron from the Iron Ranges would be crushed down into pellets of iron oxide ore called taconite, then shipped in trains down to ports on Lake Superior, especially Duluth, Minnesota. The iron boats would take their cargo across Lake Superior, down Lake Huron, and into Lake Erie to Cleveland, where it would be shipped to the steel mills located in eastern Ohio and western Pennsylvania. There, rivers (like the three rivers—Allegheny, Monongahela, and Ohio—that surround Pittsburgh) would bring barges full of coal from the Appalachian mines nearby, powering the furnaces for the smelters that turned raw taconite iron into high-grade steel.

The iron of the Iron Ranges even had a cultural impact. Canadian folksinger Gordon Lightfoot had a hit song in 1976 when he wrote about the tragic loss of the iron boat *Edmund Fitzgerald* in a storm on Lake Superior the previous year. Hibbing, Minnesota, is famous for being the home of people like baseball great Roger Maris (who broke Babe Ruth's home run record) and basketball great Kevin McHale of the champion Celtics teams in the 1980s. Folk singer Bob Dylan, born in Duluth but raised in Hibbing, described the harsh life of the iron miners he knew in his song "North Country Blues" (1963).

The resources and culture of the iron mines left a huge imprint on America, but by the 1970s and 1980s, most of the mines had closed. Cheaper iron ore was coming from many other places in the world, especially the immense iron deposits in the Hamersley Range in the Pilbara region of Western Australia. The Hamersley open-pit mines are so large that they can be seen from space and are now the largest iron producers in the world. In 2014 Australia produced 430 million metric tonnes of ore, most of it from the mines in the Hamersley Range. Some geologists estimate that Australia has 24,000 million tonnes of ore remaining in its iron ranges. By contrast, the United States only produced 58 million tonnes in 2014. However, the huge demand for steel from China in recent years has exceeded what the

Australian mines can produce, and some of the American iron mines have reopened.

BIFS, GIFS, AND LIPS

How did so much iron get concentrated in places like the Iron Ranges of Minnesota or the Hamersley Range in Australia? Most of these deposits come from what are known as banded iron formations (BIFs to a geologist). As their name suggests, these rocks have red bands of iron (figure 14.4), about a few millimeters to a centimeter thick, alternating with bands that are made of pure silica (in the form of chert or jasper). Sometimes there can be thousands of these alternating bands in a row, extending over huge areas of outcrop. When these were first discovered in the mid-1800s their meaning was a mystery. Even more surprising, the rock is made of pure iron plus chert with little or no mud or sand, which you might normally expect to find washing out into the ancient seas when the iron was being deposited.

So how did sediments consisting of dissolved iron and silica settle out on the seafloor without being mixed with sand and mud? The first thing

Figure 14.4 ▲
An outcrop of banded iron formation. (Courtesy of Wikimedia Commons)

to know is that in the present day, iron cannot stay dissolved in seawater because it is rapidly oxidized to various forms of iron oxide ("rust") and clings to other minerals or settles out. The only way to transport and concentrate huge amounts of iron in seawater is if the oxygen content is so low that iron cannot rust. This proves that the ancient seas when the iron formations were deposited must have been completely anoxic, and most geologists think the atmosphere was very low in oxygen as well.

Next, you need to have the seafloor far enough from land that almost no sand or mud from the land can mix in the deep ocean basin with the chemical deposits of iron and silica. Perhaps the iron basins were in the center of ancient seas, while the sands and muds got trapped in basins on the edge of the ancient continents. However, the Hamersley deposits seem to have formed on a shallow-marine shelf, so this model is not true of all BIFs. Finally, it would be a lot easier to deposit huge concentrations of iron if there were some abundant source of dissolved iron entering the ocean. Most geologists think the iron came largely from weathering of the basaltic lavas (which are iron rich) in the ancient mid-ocean ridges and possibly from dissolved iron produced by weathering of land rocks (which would only be possible if the rivers were completely anoxic as well). Lately, geologists working on BIFs have noticed that some of the biggest deposits occurred when the earth experienced gigantic eruptions of flood basalt, known as "large igneous provinces" (LIPs). The weathering of this excessive eruption of lava would have produced a great quantity of iron as long as the atmosphere and ocean were low enough in oxygen that iron could stay in solution and not rust.

Banded iron formations are found in some of the oldest rocks on earth, including the 3.7-billion-year-old rocks of Greenland, the Isua Supracrustals mentioned in chapters 12 and 13. Most of the world's BIFs (figure 14.5) were produced during the Archean (4.0 Ga to about 2.5 Ga), when the earth not only had an anoxic atmosphere, but also was covered by small proto-continents bashing around in proto-oceans made of a weird lava called komatiite (see chapter 12). By 2.6 Ga to 2.4 Ga, the largest volume of BIFs was deposited, especially the huge mountains of iron in the Hamersley Range of Australia, the Iron Ranges around Lake Superior, and similar deposits in Brazil, Russia, Ukraine, and South Africa. This window of time was also when the huge eruptions in the LIPs were at their peak. The BIFs began to disappear, although there were still large deposits of iron in granular form, known as GIFs (granular iron formations).

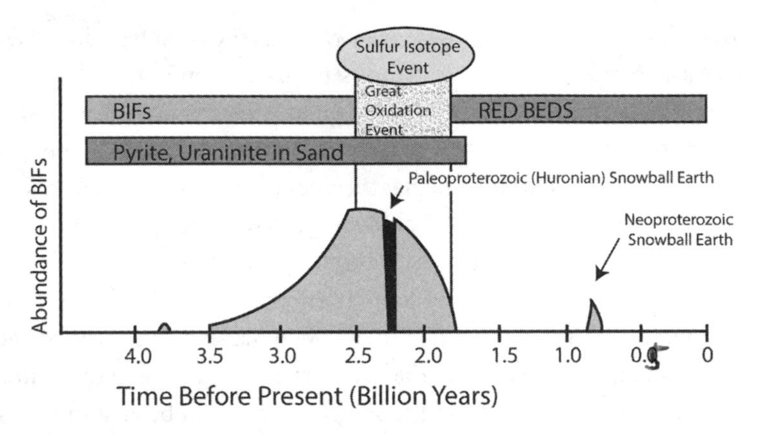

Figure 14.5 ▲

Time scale of the appearance of BIFs, stromatolites, and oxygen concentrations in the Precambrian.

THE OXYGEN HOLOCAUST

Then, around 2.3 Ga, something happened. By 1.9 Ga, BIFs and GIFs vanished completely, except for a few freak occurrences (see figure 14.5) during a "snowball earth" episode around 750–580 Ma (see chapter 16). Most geologists regard this as the time when oxygen finally began to reach significant levels in the earth's atmosphere, and possibly in the ocean as well. It has come to be called the "Great Oxidation Event" (GOE in the trade). Oxygen was still nowhere near the 21 percent that is now found in the earth's atmosphere. Instead, it went from almost nothing before 2.4 Ga to about 1 percent of the present levels in the oceans, which is enough to begin rusting the dissolved iron in the oceans. Then, about 1.9 Ga, geologists think that the oxygen levels in the oceans were high enough that oxygen was escaping to the atmosphere and possibly weathering rocks on land, although it was still not abundant in the atmosphere. It is thought that oxygen reached levels like we have today only in the last 500 million years, saturating the oceans and atmosphere so they are now completely oxygenated.

How do we know that oxygen levels were this low? The best evidence comes from the BIFs, which could only form if the oceans were so low in oxygen that iron could stay dissolved, rather than rusting and being left behind. There are also other geochemical indicators. Before 1.9–1.8 Ga, we find sand grains and pebbles in the river deposits made of the mineral pyrite or "fool's gold," which is iron sulfide (FeS_2). Today, pyrite only forms in places with very low oxygen levels, such as the bottom of stagnant bodies of water or in deep hot springs and crustal rocks far from the atmosphere. Once pyrite grains weather at the surface, they quickly break down into pebbles of iron oxide, rather than iron sulfide. I have collected specimens of iron oxide that still had the crystal form of pyrite, even though the mineralogy had changed. As the pyrite breaks down, the iron is released and the sulfur is oxidized into sulfate to make minerals like gypsum (calcium sulfate, or $CaSO_4$). Not surprisingly, we find few significant gypsum deposits older than about 1.8 Ga, and no more pyrite pebbles or sand grains after that same time. Sand grains made of uranium oxide (uraninite, UO_2) are common before 1.7 Ga, but never found after that time. Like pyrite sand and dissolved iron, they are unstable in an oxygen-rich atmosphere.

There are other indicators as well. If we look at the record of isotopes of carbon through time, we no longer see really low values associated with low oxygen after about 2.2 Ga. Sulfur isotope values in Archean rocks are highly variable, fluctuating all over the place. But after 2.4 Ga, the sulfur isotopes are highly stable, because they are no longer floating free in minerals like pyrite but are stabilized in gypsum and other minerals common in an oxygen-rich world.

Thus, the world went through a dramatic transformation as soon as oxygen became available. The GOE has also been nicknamed the "Oxygen Holocaust," because for life that was used to anoxic conditions, the appearance of such a reactive molecule as O_2 meant death (see chapter 13). Today, these bacteria and other microbes that are adapted to low-oxygen conditions must live in oxygen-starved places like the bottoms of stagnant lakes and marine basins like the Black Sea. Before 2.3 Ga, however, they ruled the planet. Once the atmosphere became too rich in oxygen, it truly was a holocaust for them, and they lost the world to microbes that can survive oxygen-rich conditions.

The burning question then arises: Where did the earth's atmosphere get its free oxygen? The answer is clear: photosynthesis, first from blue-green bacteria or cyanobacteria (see chapter 13), and then eventually when eukaryotic "true" algae evolved, from plants as well. The big puzzle is that cyanobacterial fossils are known from 3.5 Ga and possibly even 3.8 Ga, yet the GOE doesn't start to happen until about 2.3–1.9 Ga. Was cyanobacterial oxygen production so meager that it didn't make a dent on the planet? Did cyanobacteria produce lots of oxygen, but it mostly got locked into crustal rocks that were oxidized (such as BIFs) until so much oxygen was finally produced that the crustal reservoirs were saturated and there was free oxygen left over? Or maybe 2.3 Ga is when true eukaryotic algae evolved, with their much larger cells and greater oxygen production. Maybe only true algae could produce enough oxygen to overwhelm the earth's oxygen sinks, while much smaller cyanobacteria could not. Whatever the reason, the argument remains very controversial and speculative, and there is no consensus on the answer. What is clear is that after 1.7 Ga, there were true eukaryotic algae everywhere, and there was an atmosphere with about 1 percent or maybe even more oxygen, forever changing the earth's oxygen balance.

Here's another thing to think about: without some free oxygen, multicellular animals could not have evolved—and we would not be discussing the issue, since humans could never have evolved either. In fact, the evolution of all of life (except anaerobic microbes) depends on an oxygen-rich planet, which cannot happen without the evolution of photosynthetic microbes and plants. This is a severe restriction on the speculative ideas about extraterrestrials and life on other planets as well. It's true that astronomers have found lots of other planets with Earth-like properties, including the right size, the right temperatures, and possibly even liquid water oceans on their surface. But so far, not one has evidence of free oxygen in its atmosphere. Without it, there is no multicellular animal life, and no alien being like you see in so many science fiction movies (and in that entire culture of people who believe in aliens and UFOs). It's possible that there are anoxic microbes in the deep crustal rocks of other planets, but without abundant free oxygen, the aliens of other planets and our imagination don't exist.

FOR FURTHER READING

Canfield, Donald E. *Oxygen: A Four Billion Year History*. Princeton, N.J.: Princeton University Press, 2014.

Hazen, Robert M. *The Story of the Earth: The First 4.5 Billion Years from Stardust to Living Planet*. New York: Penguin, 2013.

Knoll, Andrew H. *Life on a Young Planet: The First Three Billion Years of Evolution on Earth*. Princeton, N.J.: Princeton University Press, 2003.

Lane, Nick. *Oxygen: The Molecule That Made the World*. Oxford: Oxford University Press, 2003.

Schopf, J. William. *Cradle of Life: The Discovery of Earth's Earliest Fossils*. Princeton, N.J.: Princeton University Press, 1999.

Shaw, George H. *Earth's Early Atmosphere and Oceans, and the Origin of Life*. Berlin: Springer, 2015.

Ward, Peter, and Joe Kirschvink. *A New History of Life: The Radical New Discoveries About the Origin and Evolution of Life on Earth*. New York: Bloomsbury, 2015.

15 ARCHEAN SEDIMENTS AND SUBMARINE LANDSLIDES

TURBIDITES

The sediments of the past are many miles in collective thickness: yet the feeble silt of the rivers built them all from base to summit.

—JOHN JOLY

PUZZLE PIECE #1: THE MYSTERY OF THE BROKEN CABLES

November 18, 1929—People in the small fishing communities of Newfoundland, Nova Scotia, and other parts of maritime Canada were shaken violently by a strong earthquake at 5:02 P.M. local time. The quake was felt as far as New York and Montreal. Damage was extensive throughout the region, totaling at least $400,000. Even more destructive, a series of tsunamis (seismic sea waves) rushed onto the shore and destroyed most of the coastal communities. Only two hours after the quake, a 20-foot-high tsunami struck the coast of Newfoundland, even reaching as far as Bermuda, 1,445 kilometers (900 miles) from the epicenter. Smaller waves even reached South Carolina and crossed the Atlantic to Portugal. The coastal towns of Newfoundland's Burin Peninsula were nearly wiped out. According to one source:

> The force of the waves lifted houses off their foundations, swept schooners and other vessels out to sea, destroyed stages and flakes, and damaged wharves, fish stores, and other structures along the peninsula's extensive

> coastline. Approximately 127,000 kilograms of salt cod were also washed away by the tsunami, which affected more than 40 communities on the Burin Peninsula. At Point au Gaul, giant waves destroyed close to 100 buildings as well as much of the community's fishing gear and food supplies; St. Lawrence lost all of its flakes, stages, and motorboats. Government assessment later placed property damage on the Burin Peninsula at $1 million.
>
> Worse than the damage to property, however, was the loss of human life. The tsunami killed 28 people in southern Newfoundland, which is more than any other documented earthquake-related event in Canadian history. Twenty-five victims drowned during the disaster (six bodies were washed out to sea and never found) and another three later died from shock or other tsunami-related conditions. The deaths were confined to six communities: Allan's Island, Kelly's Cove, Point au Gaul, Lord's Cove, Taylor's Bay, and Port au Bras. Fortunately, the tsunami struck on a calm evening when most people were still awake and could quickly react to the rising water; many managed to evacuate their homes and flee to higher ground.

Three days later emergency relief ships, including the the first to arrive, the SS *Meigle*, brought medicine, supplies, food, and doctors and nurses to help care for the injured and sick and get the region on the road to recovery. Another account describes the efforts of relief ships:

> Early on the morning of November 21, the SS *Portia* made a scheduled stop at Burin Harbour. Fortunately, the *Portia* had a wireless radio on board as well as an operator who immediately sent a wireless message to St. John's describing the situation. The ship's captain, Westbury Kean, later wrote to the *Evening Telegram* of his shock at seeing the damage: "Imagine our wonder and surprise on turning the point of the channel to be met by a large store drifting slowly along the shore seaward; then a short distance another store or a dwelling house until 9 buildings were counted, strewn along the shore before the harbour was reached. On reaching the harbour even a worse spectacle greeted the eyes."

Donations totaling a quarter of a million dollars came from the United States, Canada, and the United Kingdom.

People didn't know it at the time, but it was an extraordinarily strong quake for this part of the world, with a magnitude of 7.2. Quakes of that size usually cause much more damage, but this one was located not on the land beneath the towns, but 400 kilometers (250 miles) south of Newfoundland. Its epicenter was in deep water in the Grand Banks of Newfoundland, a famous fishing area that sustained the largest cod fishery in the world. Thus, it was so far offshore that the seismic waves traveled a long distance and lost a lot of energy before they reached civilization.

One other peculiar thing happened: nearly all the telegraph and telephone service between North America and Europe was cut off. Radio, satellite, microwave, and other wireless forms of communication did not exist at that time. Instead, all the telephone and telegraph communications across the Atlantic were transmitted via huge transatlantic cables that had been laid by ships starting in 1858 (figure 15.1). The cable itself was partially designed by the famous physicist Lord Kelvin, whom we met in chapter 8. When they were completed, the cables traversed 2,300 nautical miles (4,300 kilometers) across the deep bottom of the Atlantic Ocean, from England to Newfoundland, where messages were then relayed by land cables across North America. There were twelve such cables on the Atlantic continental shelf crossing the Grand Banks by 1929.

Eventually, ships were dispatched to retrieve the cables and patch them up, and transatlantic communication resumed. People certainly realized that the earthquake had broken those cables, but no one knew how. The mystery of the broken cables was filed away and forgotten over the years.

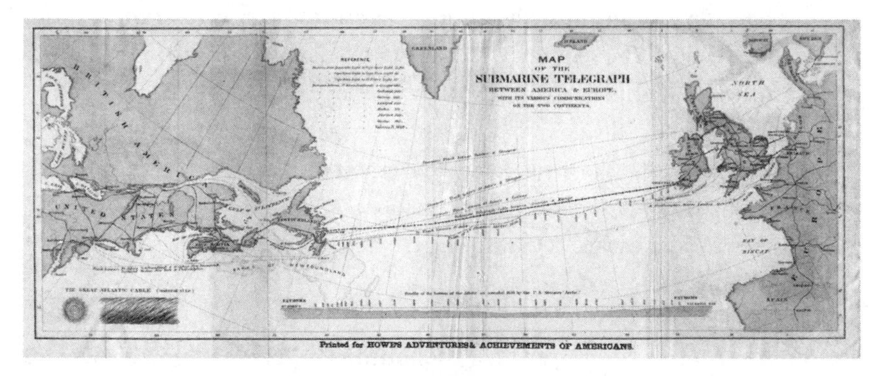

Figure 15.1 ▲

The routes of the transatlantic cable system. (Courtesy of Wikimedia Commons)

PUZZLE PIECE #2: THE MYSTERY OF THE GRADED BEDS

Geologists have long struggled to understand the rocks of the earliest earth, especially from the time known as the Archean Era (4.0–2.5 Ga). As we have seen in other chapters, they seem to indicate a world very different from anything we know today. For example, the continental crust was very thin and hot and composed of small microcontinents or protocontinents; it was not the large, thick, cold continental crust we see today. The oceanic floor was formed of lavas that erupted from deep in lowermost crust, as well as the mantle, and their composition was a peculiar olivine-rich lava known as komatiite that hasn't erupted since the Precambrian—all modern ocean floor lavas are basalt. Banded iron formations (see chapter 14) are found in many of the ocean basins around the world, suggesting that the atmosphere had almost no free oxygen until after the Archean ended. And as we saw in chapter 11, the moon was much closer to Earth, so it not only appeared huge in the sky but also had a strong tidal pull, generating giant "tidal waves" that swept across the shallow seas of the globe every few hours.

Even more peculiar were the sedimentary rocks, such as sandstones. In nearly all rocks younger than the Archean, the most common sandstones are largely made of the durable mineral quartz, the most common mineral on the earth's surface. This is because most other minerals (such as feldspars, which make up the bulk of most kinds of igneous rocks) are broken down quickly by chemical weathering. Quartz, on the other hand, is chemically inert (it's just silicon dioxide, SiO_2), plus it has no cleavages to break it into fragments, as feldspars do. Consequently, when quartz grains erode out of bedrock and wash down rivers and streams, they survive all the battering and bashing against other hard grains and resist dissolving as well. Eventually, when those rivers transport their sand all the way to the lower floodplains, or even out to the beaches and the sea, most of the sand is rich in quartz, with only minor components of other minerals occurring in most beach and river sands.

Then those same sands get recycled: buried in a sedimentary basin, cemented into sandstones, uplifted again, and eroded out of mountains to form sediment a second time. Each time around, the sandstones get even richer in quartz, and nearly all the unstable sand grains (such as those from feldspars or rock fragments) are completely eradicated. The higher the percentage of quartz and other very durable minerals (such as zircon, tourmaline, and rutile, or the ZTR index—see chapter 12), the more "mature" a sandstone is. Some sandstones are considered "supermature" because they are 99.99 percent pure quartz, with extremely well-rounded and well-sorted grains, all of the same size. Sedimentary geologists have long puzzled over what conditions might push a quartz sand to this extreme. Most of them now agree that to get a sand this rich in quartz and well-rounded grains, the sand grains must have spent at least part of their time in a wind-blown sand dune before finally being dumped back into the ocean for the last time and cemented into sandstone.

This was what sedimentary geologists had come to expect for any sandstone less than 2 billion years old, and this is how we see sandstones forming today. But when geologists went to those rare places where Archean sedimentary rocks were exposed (such as the center of Canada or South Africa or Brazil), they were surprised. There were no normal sandstones at all. Instead, almost all the sedimentary rocks (other than banded iron formations) were thick, parallel sandstone beds alternating with shale beds dozens to hundreds of times in a row and extending sideways for long distances without changing thickness (figure 15.2).

Another peculiarity was the nature of these sandstones. They were not clean pure quartz sandstones so typical of later times, but a rock the German geologists called *grauwacke*—an immature sandstone with lots of mud trapped between the grains, rather than the clear pore space found in normal sandstones. Even more surprising, each graywacke sandstone layer showed a peculiar feature known as graded bedding (figure 15.3), where the base of the bed was the coarser material (gravel and coarse sand), and the sand became finer and finer as you traveled to the top of the bed, often with the very top of the unit composed of fine-grained shales.

All of these features were peculiar, although there were some places in the world (such as basins in front of the rising Alps in Germany or deep-marine basins in Southern California) where very young graywackes with graded bedding alternating with shales were known. But it was still a mystery how they formed. The graded bedding suggested a mixture of sand and mud that had settled out of suspension, with the largest grains sinking fastest, and the fine-grained muds sinking very slowly, so the grain sizes were graded from coarse to fine. That was all clear enough, but how to explain why it had happened hundreds of times in a row? Geologists could imagine

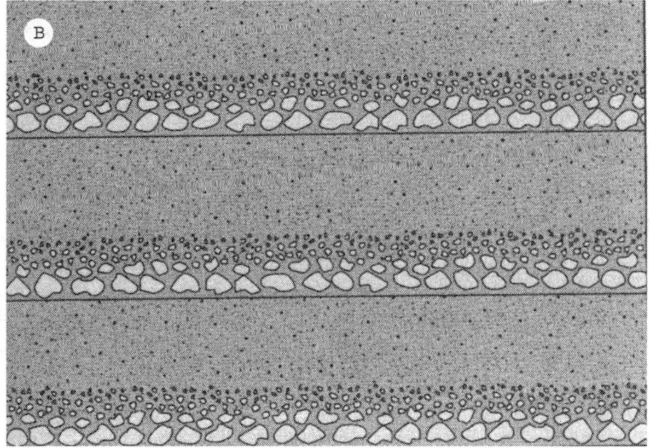

Figure 15.2 ▲

(*A*) Repetitive graded beds, typical of Archean sedimentary deposits. (*B*) Diagram of repetitive graded beds, showing the coarse material at the base and the finer material at the top of each bed. ([*A*] Courtesy of Karl Wirth; [*B*] Courtesy of Wikimedia Commons)

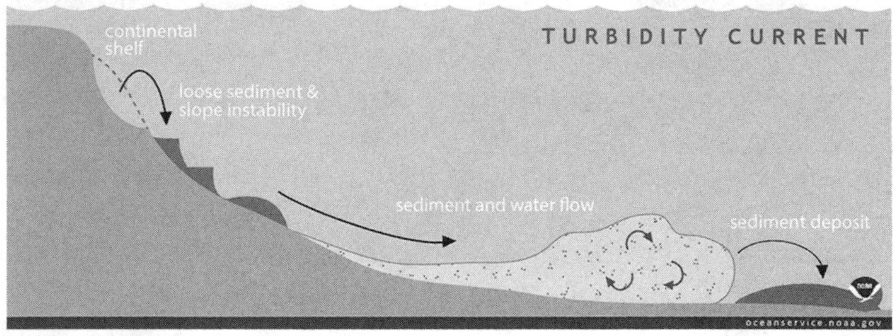

Figure 15.3 ▲

The mechanism of huge gravity flows known as turbidity currents that are responsible for the formation of graded bedding. (Courtesy of NOAA)

a single graded bed representing a transition from shallow-water sandstone to deepwater shales, but to make it happen hundreds of times in a row, sea level would have to go up and down rapidly like a yo-yo—a geological absurdity.

The more geologists puzzled over these peculiar rocks, the more confusing they became. The legendary sedimentary geologist Francis J. Pettijohn describes his mapping in the Archean of northern Canada in the late 1920s and early 1930s in his famous 1984 biography *Memoirs of an Unrepentant Field Geologist*. As he wrote (p. 123):

> I was struck by the prevalence of graywacke. All the Archean sandstones were graywackes—dark rocks filled with angular quartz, feldspar, and rock particles. Why were the Archean sandstones so different in aspect from the clean white Precambrian quartzites of the Huronian [early Proterozoic] of the north shore of Lake Huron and the quartzites of the Michigan iron ranges? Moreover, the Archean assemblage was distinctive—greenstones and graywackes, with no limestones or quartz arenites [sandstones].

Despite their puzzlement, geologists tried their best to explain these peculiar deposits. In 1930 E. B. Bailey suggested they were caused by periodic earthquakes disturbing the sediment, which then settled out slowly afterward. Others had even more imaginative mechanisms. Most simply settled for describing them in detail but refrained from speculation about their causes. What was clear about Archean sandstones was that they were all immature graywackes, clearly freshly eroded from the land without any sorting, winnowing, or recycling, which explained the absence of clean quartz sandstone. They were clearly deposits formed by settling of coarser and then finer grains by gravity, but how this happened and why it was so repetitive and rhythmic was still a mystery.

PUZZLE PIECE #3: TURBIDITY CURRENTS

In a completely different area of science, some important discoveries were being made. When Hoover Dam was completed in 1936 and Lake Mead began to fill up in the late 1930s, engineers found to their surprise that not all the sand sinking out of the Colorado River and being dumped at the head of the reservoir was staying there. When they took samples much farther down from the river's outflow, they found thick sand beds that had flowed across the bottom of the still water in the reservoir, sometimes for hundreds of miles, with no currents to aid them. The engineers actually measured sedimentary flows sliding along the bottom of the reservoir at speeds of up to 30 centimeters per second (0.6 miles per hour). These bodies of flowing sand were denser than water (1.05 grams per cubic centimeter, rather than the 1.0 grams per cubic centimeter found in clear water) and formed sand beds with grading up to 2 meters (6.5 feet) thick. In 1936, a geologist named Daly suggested that these sandy gravity flows (now called turbidity currents) found on the bottom of Lake Mead might also explain the graded beds found in other places, but there were no experimental data to confirm this idea.

What geologists needed were simulations and experiments that showed exactly how these submarine landslides, or turbidity currents, worked. Into this breach stepped an innovative Dutch geologist by the name of Philip Kuenen at Groeningen University. Before World War II, he had been on an oceanographic survey on the research ship *Snellius* to the Dutch East Indies in 1929–1930. On that voyage, the ship found evidence of huge submarine canyons eroded into the continental shelf, which Kuenen published a paper on in 1937. The voyage had dredged up sands from the deep ocean and had taken cores of the bottom that seemed to show graded beds forming in deep water. Based on these data, Kuenen published a paper in 1938 speculating that gravity-driven currents (essentially, submarine landslides) had eroded these submarine canyons and deposited the graded beds of sandstone.

Once World War II was over, Kuenen decided to test his ideas. He built long narrow flumes that were about a foot wide and many feet long in his lab area and filled them with water. Some flumes had clear glass sides so you could see through the water inside and watch the currents move past any spot. The flumes had a slight slope at one end, but there were no currents running in them. In essence, they were like one extremely long but shallow and narrow fish tank. Then he dumped a mix of sand and mud into one end and watched it flow down the flume by gravity alone. At the beginning, it was a large, cloudy, turbulent suspension of mud and sand, churning and swirling around as it was freshly dumped into the water. But soon it became a turbulent flow, quickly oozing down the bottom of the flume as a separate body of water, separated from the clear water above by its density and not mixing with the clear water (figure 15.4). His experiments

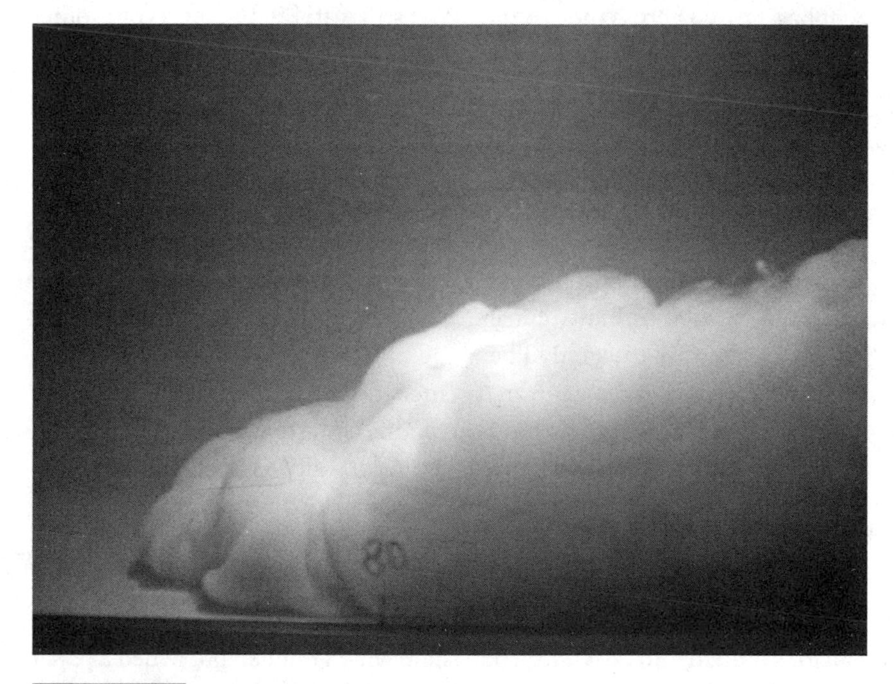

Figure 15.4 ▲

Photograph of natural underwater turbidity current, showing the dense turbulent mass of suspended sediment that flows beneath the clear water above it. The turbidity current does not mix with the clear water, but remains separate and distinct. (Courtesy of Wikimedia Commons)

were published in a paper with Italian geologist C. I. Migliorini in 1950, and another paper by Kuenen alone in 1951. Soon, geologists were looking at turbidites all over the world, and the mysterious graded beds of the Archean began to make sense.

PUZZLE SOLVED!

The turbidity current model was suggested by the samples dredged and cored from deep water, from outcrops of ancient graded beds, and from Kuenen's experiments. But since they happened in very deep water (over a mile deep), no one could watch them occur in real time.

Then a pioneering oceanographer named Bruce Heezen (pronounced "HAY-zen") at Lamont-Doherty Geological Observatory (now Lamont-Doherty Earth Observatory) of Columbia University had a flash of inspiration. (I knew Bruce and his partner Marie Tharp, who mapped the entire ocean floor, when I was at Lamont and Columbia in the late 1970s before Bruce died of a heart attack while doing research in a submarine in 1977.) Bruce was working on his data from the seafloor of the Grand Banks, when he stumbled across accounts of the 1929 Grand Banks earthquake and how 12 different transatlantic cables had abruptly broken, cutting off telegraph and telephone service across the ocean. Looking closer, it was possible to know exactly when each cable had broken, because the precise time of the telegraph service interruption was known. He then plotted when each cable was broken and where they were on the surface of the Grand Banks or the slope of the continental rise (the gently sloping region at the base of the continental slope) down to the deep ocean. Sure enough, they formed a sequence, with the first cables broken near the top of the continental shelf, and each subsequent break occurring in cables farther down the slope to the deep ocean. Could these cables have acted as inadvertent trip wires, breaking in sequence as a powerful sandy turbidity current triggered by the earthquake roared down from the shallow shelf? All it took was to calculate the time the break occurred after the earthquake and the distance of the trip wire/cable from the epicenter and then plot them (figure 15.5). Sure enough, they formed a smooth plot, and the slope of the line of the plot gave the velocity at different stretches of the submarine landslide as it swept 600 kilometers (400 miles) down the shallow shelf to the deep

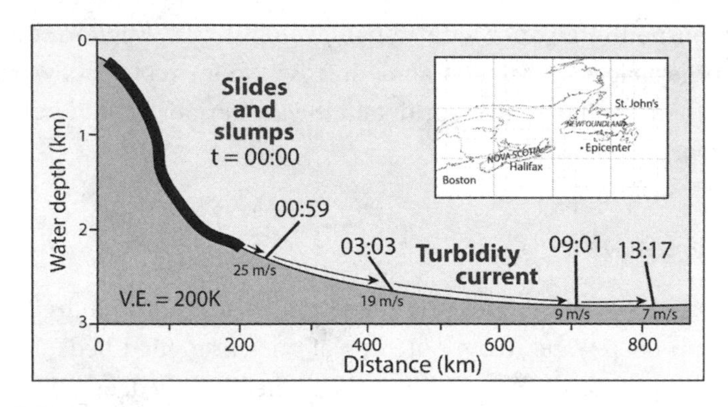

Figure 15.5 ▲

Diagram of the events of the 1929 Grand Banks earthquake and its speed as it broke 12 transatlantic cables. (Redrawn from several sources)

oceanic abyss. At the top, where the slope was steepest, the submarine landslide was moving at 25 meters per second (55 miles per hour). As the current slowed in deeper water, where the slope became gradual, it dropped to 9 meters per second (20 miles per hour) and then 7 meters per second (15 miles per hour). It was still moving at speeds nearly as fast for many more hours, even when the ocean floor had no slope at all, because the gravity slide had a huge momentum behind it, so it could flow along the flat ocean floor for many miles.

In 1952 Heezen wrote up this amazing natural experiment in geology and published it with his boss, the founder of Lamont, Maurice "Doc" Ewing. Finally, the mystery of the mysterious graded beds from the Archean rocks that were more than 2.5 billion years old had been solved by a determined Dutch experimentalist—and by an accidental experiment that nature had provided due to a deadly earthquake. Science sometimes works in remarkable and mysterious ways.

FOR FURTHER READING

Bouma, Arnold. *Turbidites*. Springer, Berlin, 1964.

Bouma, Arnold H., and Aart Brouwer, eds. *Turbidites*. Amsterdam: Elsevier, 1964.

Bouma, Arnold H., William R. Normark, and Neal E. Barnes, eds. *Submarine Fans and Related Turbidite Systems*. Berlin: Springer, 1984.

Bouma, Arnold H., and Charles G. Stone. *Fine-Grained Turbidite Systems*. Tulsa, Okla.: American Association of Petroleum Geologists, 2000.

Pettijohn, F. J. *Memoirs of an Unrepentant Field Geologist: A Candid Profile of Some Geologists and Their Science, 1921–1981*. Chicago: University of Chicago Press, 1984.

Weimer, Paul, and Martin H. Link, eds. *Seismic Facies and Sedimentary Processes of Submarine Fans and Turbidite Systems*. Berlin: Springer, 1991.

16 | TROPICAL GLACIERS AND THE SNOWBALL EARTH

DIAMICTITES

I think about the cosmic snowball theory. A few million years from now the sun will burn out and lose its gravitational pull. The earth will turn into a giant snowball and be hurled through space. When that happens it won't matter if I get this guy out.

—BASEBALL PITCHER BILL LEE

PUZZLE DOWN UNDER

Doing geology in Australia can be both a blessing and a challenge. On the plus side, most of the continent is a dry desert or scrub habitat, so there are lots of bare, exposed outcrops of rock. There is very little vegetation, unlike wetter parts of the world, where the plants overgrowing nearly everything hampered early British geologists (chapters 4–7) and the other geologists who came after them. Most of my geological career has been spent in deserts and badlands, because these are the only places with suitable exposures of beds to find fossils. On the minus side, Australia is a large, stable block of continental crust that has undergone relatively few great collisions or mountain-building events that force sedimentary basins to sink and rocks to erode, and none since 250 Ma. Most of the sedimentary deposits in much of Australia are very thin and discontinuous, so they don't lend themselves to the study of long sequences of rocks and fossils through time, unlike many other parts of the world.

Some parts of the geological record are missing or nearly so. Australia has a thick sequence of deposits for much of the Precambrian (such as the stromatolites in chapter 13, the banded iron formations in chapter 14), especially the upper Precambrian rocks that are the subject of this chapter. Most of the Paleozoic units are relatively thin and unfossiliferous compared with many other continents. The Australian Mesozoic record has some bright spots, but no thick stack of richly fossiliferous dinosaur-bearing beds like you find in North and South America and Eurasia and Africa. By the Cenozoic, tectonic activity shut down almost completely, so there are very few mammal-bearing beds and a relatively poor fossil record of the Cenozoic. The exceptions to this rule are unusual deposits like Riversleigh, from the Miocene of Queensland, which consists mostly of fossil mammals and other terrestrial animals that fell into sinkholes in limestone.

Nevertheless, Australian geologists have made the best of what they have. The Precambrian banded iron formations have been intensively studied by many scientists, as have the late Proterozoic soft-bodied fossils of the Flinders Range, or the incredible beautifully preserved Devonian fish of the Gogo Formation.

One of these geologists was the legendary Sir Douglas Mawson (1882–1958), who was also a famous explorer (figure 16.1). He began his career mapping rocks in Melanesia and then in New South Wales, but in 1907 he joined Sir Ernest Shackleton's expedition to Antarctica, where he stayed for over 2 years, after most of the rest of the explorers had returned. Mawson was the first man to reach the top of Antarctica's second highest volcano, Mount Erebus, at 3,794 meters (12,448 feet). He also was one of the first to reach the South Magnetic Pole. In a fortunate twist of fate, he turned down a chance to join Robert Falcon Scott's ill-fated 1910 expedition to the South Pole. They reached their destination only after Norwegian Roald Amundsen had gotten there first, then all of Scott's crew died trying to return.

Instead, Mawson led his own Australasian Antarctic Expedition in 1911, which mapped and studied much of East Antarctica. Despite great discoveries, the conditions got bad and most of the explorers perished on their return to home base. After everyone else died, Mawson and his last companion, Xavier Mertz, ate all their sled dogs. They both got hypervitaminosis A from eating too much dog liver, and Mertz died. Alone, Mawson hiked back over hundreds of miles in subfreezing conditions, even falling into a crevasse and dangling over the abyss in the sledge harness. Eventually,

A

B

inspired by a line of poetry, he managed to pull himself out. As conditions got worse, the soles of his feet were so frostbitten that they detached from the flesh beneath. Then, just hours before he reached home base, his rescue ship sailed away. Even though they were called back by radio, bad weather prevented their return for days. When Mawson was finally rescued, he had been trapped in Antarctica for almost 3 years. The story is told in David Roberts's book *Alone on the Ice: The Greatest Survival Story in the History of Exploration.* In his own book, *Home of the Blizzard*, Mawson recounts the entire grueling experience, including surviving subfreezing winds on Cape Denison that averaged 60 miles per hour, with gusts that approached 200 miles per hour.

After recovered from this harrowing experience, Mawson later helped with the search to retrieve the frozen corpses and journals of the ill-fated Scott expedition. He served in the British Army in World War I, and then returned to Australia in 1919 to become a professor of geology at the University of Adelaide until his retirement in 1952. He spent most of his career mapping and studying geology in Australia, especially the Flinders Range in South Australia, now famous for its upper Proterozoic rocks and the earliest megascopic body fossils of the strange creatures known as the Ediacara fauna. Still, Antarctic exploration was in his blood, and in 1929–1931 he led the joint British Australian and New Zealand Antarctic Expedition, which resulted in the formation of the Australian Antarctic Territory. Mawson died at age 76 in 1958, not in some frozen wasteland, but quietly in his bed of old age. He is so famous and so venerated as one of Australia's greatest explorers and scientists that he appeared on the Australian 100 dollar bill, and his name is on many landmarks in both Australia and Antarctica.

It was lucky that Mawson was so familiar with ice sheets and glacial deposits through his years of experience on the Antarctic continent. In the upper Precambrian rocks of the Flinders Ranges and elsewhere in South Australia, he found thick deposits of what is known as glacial till (figure 16.2). Till is an unsorted, unstratified mass of boulders, gravels, sand, and mud

Figure 16.1 ◀

Photographs of Douglas Mawson: (*A*) Mawson resting on a sledge at the beginning of the Australasian Antarctic Expedition to Antarctica in 1912. (*B*) Photo taken in 1913, with a bearded Mawson still suffering from frostbite and malnutrition after he returned as the sole survivor of the expedition. (Courtesy of Wikimedia Commons)

Figure 16.2 ▲
Photograph of the glacial tills of the late Precambrian Elatina diamictite of Australia. (Courtesy of Paul Hoffman)

that is randomly dumped by the snout of the glacier as it melts in one place. These sediments are very distinctive, and almost nothing else produces anything like them, so ancient glaciation can be recognized in the rock record. However, many geologists prefer to use the term "diamictite" (meaning "thoroughly mixed" in Greek) or sometimes the word "tilloid" as a noncommittal way of describing a rock with this texture, without directly implying that is it glacial. Following work by Oskar Kulling in 1934 and by Walter Howchin, Mawson was eventually convinced that this was evidence of a global glacial event in the late Proterozoic, because the ancient glacial deposits in Australia were not far from the modern equator. But by the late 1950s and early 1960s, geologists began to dismiss his argument, because plate tectonics had shown that Australia and other continents had moved long distances over time. Conceivably, the Australian Precambrian glacial beds could be from a time when the continent was closer to the South Pole. Ironically, we now know that Australia was right on the equator at that time, even more tropical than Mawson thought, so his evidence was stronger than anyone realized.

THE ICEBED-LIMESTONE SANDWICH

But the idea of a global late Precambrian glaciation wouldn't die. In 1964 Cambridge geologist W. Brian Harland (1917–2003) published a famous paper that showed that tropical upper Precambrian glacial deposits were not restricted to Australia. Like Mawson, Harland was also familiar with glaciers and ice sheets firsthand, since over his lifetime he spent a lot of time in the Arctic. He established the Cambridge Arctic Shelf Program. He was part of an amazing 43 polar field seasons (of which he led 29) from 1938 to the 1960s, mapping the geology of the archipelago of Svalbard (= Spitzbergen), a group of islands between Norway and Greenland. There he saw not only the deposits of recently melted glaciers, but also upper Precambrian glacial deposits in abundance, not only on Svalbard but also in Greenland and Norway.

Harland provided more than just evidence of glaciers. He pointed out that many of the upper Precambrian ice deposits were sandwiched between layers of limestone. This is much more surprising, because today limestones are only formed in warm tropical or subtropical shallow-marine settings like the Bahamas, Florida, Yucatan, the Persian Gulf, and the South Pacific. If this icebed-limestone sandwich had been formed by modern processes, then the glacial deposit surrounded by limestones had to form in the tropics and at sea level. Today there are a few places where tropical glaciers are known, such as the top of Mount Kilimanjaro in Kenya and the Peruvian Andes, but those glaciers are in high mountains. It seems impossible to imagine sea-level tropical glaciers, but Harland's evidence was inescapable. If the tropics were glaciated, so were the poles, and so was the whole planet in the late Precambrian.

In addition, Harland backed up his conclusions with a new line of evidence: paleomagnetism. He was one of the first to measure the ancient magnetic directions frozen in the rocks since their formation, which can tell us at what latitude a given rock unit was formed. The paleomagnetic directions for all these Svalbard, Greenland, and Norwegian rocks were tropical or subtropical in the Precambrian, so the limestone-till-limestone sandwich was not some fluke. The paleomagnetic data from Australia was not very good then, but later analyses showed that Mawson's Precambrian icebed-limestone sandwiches were right on the equator. Clearly, something weird was going on if there was indisputable evidence of tropical sea-level ice at that time.

THE SNOWBALL IS FORMED

In the 1960s and 1970s, many geologists were still not sure what to make of Mawson's and Harland's data and arguments, since it seemed inconceivable that the earth could have frozen all the way to the equator. Despite the evidence, they tended to dismiss this conclusion, because many were not confident that the paleomagnetic data were reliable. In addition, they could visualize less extreme scenarios where individual local regions might shift from limestone to glacial beds without freezing the whole world. The greatest difficulty of all, however, was imagining how the earth could freeze so completely. How could it flip so rapidly from a warm tropical limestone world to a tropical glaciated world, and back from glacial deposits to tropical limestones?

The answer to this came from a surprising direction: climate modeling. In 1969, Russian geophysicist Mikhail Budyko of the Leningrad Geophysical Observatory published a paper that showed how easy it is for a planet to freeze over once the ice sheets start to grow. He pointed to a well-known climate effect known as the albedo feedback loop. "Albedo" is just a fancy word for describing the reflectivity of a surface. As you know if you've ever spent time skiing or snowboarding, snow or an ice sheet has high albedo, since it reflects most of the sunlight that hits it. That's why you need good dark goggles that reduce glare with tinted lenses when you spend time on the ice. By contrast, dark surfaces (like forests or the open ocean) absorb a lot more sunlight and reflect very little.

The albedo feedback system is very sensitive to small changes, which can transform it from frozen to ice-free and back to frozen very quickly. For example, let's say the earth's surface is covered by ice, so it has a high albedo and reflects most of the sun's energy back. But the planet begins to warm slightly, and that ice sheet melts back a bit, exposing dark land and water. This absorbs more sunlight and generates heat, which melts the ice even further. Back and forth these two processes go in a feedback loop that eventually melts the ice in a very short time. Now let's imagine this dark land and ocean surface has a few really cold winters and the reflective snow and ice layer lasts a bit longer. The increased ice cover reflects more energy back out to space, and the land gets colder, so even more ice sticks around the next few winters, and the ice sheet expands. Before you know it, the entire system has switched back into a complete ice age.

Although scientists knew albedo was a key feature of the polar regions and explained why they were so sensitive to small changes in global temperature, Budyko went further. In what he dubbed an "ice catastrophe," he showed that if you had even a small ice sheet in the subtropical or tropical latitudes to start, the albedo feedback loop would kick into high gear, and the entire planet could freeze over rapidly. The only dilemma with this model was how to thaw the planet once it is completely frozen and has such a high albedo that most of its energy is reflected back to space. A completely frozen reflective iceball is a dead end, and the warming part of the feedback loop cannot rescue it.

The solution was first suggested in a 1981 paper by James Walker, Paul Hays, and James Kasting. They were focused mostly on the way in which weathering of silicate minerals in soils in the landscape can absorb carbon dioxide, but in the last paragraphs of the paper, they talked about Budyko's models and how an ice cap would shut down the weathering mechanism and lead to Budyko's "ice catastrophe." In a brief sentence in the last paragraphs, they suggested another possible mechanism: volcanoes. The earth is unlike any other frozen planet in space (such as Mars and many others that have been found) in that we have an active crust with plate tectonics that powers lots of volcanoes. Volcanic eruptions release lots of gases, especially greenhouse gases like carbon dioxide, water, methane, and sulfur dioxide. If the earth were indeed completely frozen, the volcanic gases would eventually build up and warm the planet through the greenhouse effect, so the ice would finally begin to melt. And once enough dark surface had been exposed, the albedo feedback loop could kick into high gear and quickly melt from a frozen planet to an ice-free subtropical planet with limestones in the tropics.

The idea was in print and discussed by the few people working on the late Precambrian ice deposits, but still not widely known. This all changed with a legendary paper by Joe Kirschvink, now the Nico and Marilyn Van Wingen Professor of Geobiology at Caltech. Joe is one of the most brilliant people I have ever met, and he has more great ideas in a single week than most people have in a lifetime. He is one of the world's best paleomagnetists, plus he does research on all sorts of problems on the boundary between geology and biology, from magnetic bacteria and butterflies and human biomagnetism, magnetofossils, biomineralization, to innovative ideas about the Cambrian explosion, to climate change and geochemical modeling, to polar wander and reconstructing ancient continental positions. In addition, Joe designs, builds, and maintains his own lab equipment and even writes his own computer programs. Plus, he is an outstanding, provocative, mind-expanding teacher, who challenges his brilliant students at Caltech to push the boundaries. He has won the Feynman Prize for teaching at Caltech and the William Gilbert Award in Paleomagnetism from the American Geophysical Union and has even had an asteroid named after him.

In 1989 Joe put his mind to the problem of how the snowball earth escapes being totally frozen over, and revived to the solution proposed by Walker and Kasting. They were all part of the PPRG (Precambrian Paleobiology Research Group), organized by my good friend Bill Schopf of the University of California–Los Angeles. They held a big PPRG meeting in 1989, where Joe not only revived the volcanic solution to a frozen-over earth, but pointed to evidence from Mawson's Elatina Formation in Australia that it had actually occurred (see figure 16.2). Most importantly, he coined the phrase "snowball earth," which was catchy and memorable and shifted the focus from soil weathering to volcanoes and a frozen planet.

Joe has one of the world's best paleomagnetic labs at his disposal, so he did new analyses of the ancient latitudes of these Proterozoic icebed-limestone sandwiches and showed that many of them (especially Mawson's Elatina sequence in Australia) were tropical or subtropical. Then he wrote up these ideas in a short paper tucked away in a huge expensive symposium volume from the 1989 PPRG meeting about Precambrian life. After many delays, it was eventually published in 1992 (and it was so expensive that few people owned it or read it), and Joe went on to other problems. Most people would have published such a groundbreaking paper in *Nature* or *Science*, but Joe doesn't need the glory. He has so many great ideas all the time that he doesn't need to spend all his time promoting each one for long. The idea of a snowball earth was formally named and proposed, with a clear mechanism for how to make it work. Other geologists soon caught on.

THE SNOWBALL BEGINS TO GROW

At Harvard University, geologist Paul Hoffman ran into Kirschvink at the International Geological Congress in Washington, D.C., in 1989 and learned about his snowball earth idea. (I was at that meeting too, but

I was focused on other research problems.) When Hoffman began to work upper Precambrian glacial deposits in Namibia in 1993, he realized the importance of the snowball earth hypothesis, and eventually in 1997 he began to really promote it as the Namibian glacial deposits began to dominate his research.

Hoffman is a tall, gaunt, bearded, athletic field geologist who prefers to spend most of his time trekking the Canadian Arctic or the Namibian or Australian desert, looking for outcrops. He's a serious cross-country runner, so he loves hiking. He was familiar with a variety of examples of icebed-limestone sandwiches in the Proterozoic of Canada, having spent much of his career mapping the protocontinents that made up the Archean precursors of Canada. These were later assembled into the Proterozoic core that was the nucleus around which North America grew. Hoffman recruited a number of brilliant colleagues with skills in geochemistry (such as Dan Schrag), and soon they set the geological community on fire with talk about the snowball earth.

Hoffman, Schrag, and other geologists analyzed some beautifully exposed icebed-limestone sandwiches, such as those in the deserts of modern Namibia in southwestern Africa (figure 16.3). There, the limestones on top of the glacial till are particularly thick and well developed, and they showed some peculiar geochemical and mineralogical characteristics. Hoffman and Schrag suggested that these "cap carbonates" that sit on top of the glacial tills are products of inorganic precipitation of limestones once the ocean geochemistry, saturated with dissolved carbonate, was released from the grip of the ice. They are clearly not the normal kind of limestones formed today, which are precipitated by organic activity. Modern limestones are made largely by the shells of corals, molluscs, and other marine creatures, as well as calcareous algae.

Another suggestive piece of evidence is the brief return of banded iron formations (BIFs) during the peak of the late Proterozoic snowball conditions. Kirschvink pointed out that this would make sense if the earth were frozen over, because it would shut down the oceans and make them anoxic and saturated with dissolved carbonate, so they would become highly acidic oceans (as is happening to our oceans now thanks to greenhouse gases). Without runoff from the sediment flowing down rivers (now completely frozen), the sulfate input to the oceans would be shut off, and this would result in abundant dissolved iron in these acidic, low-oxygen,

Figure 16.3 ▲
Geologists Dan Schrag (*left*) and Paul Hoffman (*right*) standing on the thick glacial Ghaub Formation full of glacial boulders in Otavi, Namibia, pointing to the boundary between the glacial bed and the overlying "cap carbonate." (Courtesy of Paul Hoffman)

low-sulfur oceans. Under these conditions, iron could accumulate on the bottom as it did between 3.7 and 1.7 Ga (see figure 14.5).

So the main snowball earth model runs like this: something causes the planet to begin to cool down dramatically until large ice sheets begin to form. At that time, without abundant and complex life (as we have now) to regulate the carbon cycle and keep pumping carbon dioxide into the atmosphere (chapter 6), the planet would begin to experience a runaway albedo feedback loop and eventually freeze over from the poles to the equator. Once it was a frozen snowball, it would be stuck in that state for millions of years, just like Mars is completely frozen now after having had liquid surface water with oceans and rivers in the past. Oceanic circulation would shut down, BIFs would accumulate on the anoxic seafloor, and lots of carbon would be frozen into little cages of ice known as methane hydrates in the seafloor sediments. If nothing else happened, the earth would have stayed frozen, and we would not be here.

Unlike Mars or any other planet, however, Earth has plate tectonics and volcanoes that over a long time erupted enough greenhouse gases to finally begin to warm the planet. Once that occurred, another runaway albedo feedback loop kicked in, and the ice melted rapidly, until it was almost all gone. The carbon caged in ice in methane hydrates on the seafloor released huge amounts of methane, further accelerating the global warming. The ocean geochemistry was now so rich in dissolved carbonate that huge deposits of calcite precipitated directly out of seawater to form cap carbonates. Finally, the planet was stable again with warm tropics and cooler poles.

Further research revealed that there were at least two or three separate events in the late Proterozoic, and one in the early Proterozoic (about 2 Ga), known as the Huronian (figure 16.4). It was based on the well-known Gowganda tillite on the shores of Lake Huron and showed that snowball earth conditions are not unique but can happen multiple times if the conditions are right.

Figure 16.4 ▲

Early Proterozoic (Huronian) glacial deposits from the Gowganda till, near Lake Huron in Canada. These represent an early snowball episode, about 2.1 Ga. (Courtesy of Wikimedia Commons)

SNOWBALL OR SLUSHBALL?

Geologists, like all scientists, are naturally skeptical of new ideas, especially those that seem beyond the norm. Over the past 25 years, the snowball earth model has piled up an increasing volume of data, so most of the geological community has no choice but to accept the obvious conclusion that something like a snowball earth must have happened at least three or four times.

Still, there are dissenters. A number of geologists accept that there were equatorial sea-level glaciers in the late Proterozoic, but not that the entire tropical region had frozen over so the earth was a frozen snowball. They prefer a slightly less extreme idea, nicknamed the "slushball." In this model, there was some glaciation on the equator (the data demand it), but much of the tropical region was cold but ice-free. They point to geological evidence of sediments that could only be formed in water, not ice. However, even Kirschvink's original model allowed for some ice-free regions in the tropics, so this is not new. Also, many geologists see evidence that the snowball earth episodes had rapid fluctuations of glacial-interglacial cycles, as the most recent Ice Ages did (chapter 25), so this allows for both glacial sediments and sediments formed in running water and unfrozen oceans. Most importantly, the dating of the separate snowball events in the latest Proterozoic shows that they were globally synchronous and occurred from pole to equator at the same time. This favors a more extreme snowball earth rather than a slushball, because in slushball models the ice lines retreat when the carbon dioxide levels go up—but that is not what we see in the late Proterozoic snowball model.

One of the implications of the snowball earth model is that the late Proterozoic deep freeze apparently caused a big change in life. Before the snowball, we find abundant fossils of the spores of eukaryotic algae (known as acritarchs) in the marine sedimentary record. Then they apparently went through a mass extinction, because after the snowball earth ends, most of the diversity of acritarchs is gone. Instead, in the latest Proterozoic we see evidence of life returning to the earth with much more complex forms, including many that were multicellular. By the end of the Proterozoic, the first large multicellular animals appear all around the earth. Known as the Ediacara biota, they were first discovered by Reg Sprigg in Mawson's Flinders Ranges of Australia, then described by paleontologist Martin

Glaessner. Once these creatures flourished, the diversification of multicellular animal life (such as trilobites) launched into high gear. It has the misleading name "Cambrian explosion" but it's more like a "Cambrian long fuse," since it took 70 million years from the earliest Ediacaran creatures to the earliest trilobite.

Most fitting of all, one of the strange jellyfish-like impressions from the Ediacara fauna is named *Mawsonites spriggi*—after the two men whose mapping of the Flinders Ranges brought the Ediacara fauna to light.

FOR FURTHER READING

Hazen, Robert M. *The Story of the Earth: The First 4.5 Billion Years from Stardust to Living Planet*. New York: Penguin, 2013.

Macdougall, Doug. *Frozen Earth: The Once and Future Story of Ice Ages*. Berkeley: University of California Press, 2013.

Mawson, Douglas. *Home of the Blizzard: A Heroic Tales of Antarctic Exploration and Survival*. New York: Skyhorse, 2013.

Roberts, David. *Alone on the Ice: The Greatest Survival Story in the History of Explorations*. New York: Norton, 2014.

Schopf, J. William. *Cradle of Life: The Discovery of Earth's Earliest Fossils*. Princeton, N.J.: Princeton University Press, 1999.

Schopf, J. W., and Cornelis Klein, eds. *The Proterozoic Biosphere: A Multidisciplinary Study*. Cambridge: Cambridge University Press, 1992.

Shaw, George H. *Earth's Early Atmosphere and Oceans, and the Origin of Life*. Berlin: Springer, 2015.

Walker, Gabrielle. *Snowball Earth: The Story of a Maverick Scientist and His Theory of Global Catastrophe That Spawned Life as We Know It*. New York: Broadway, 2004.

Ward, Peter, and Joe Kirschvink. *A New History of Life: The Radical New Discoveries About the Origin and Evolution of Life on Earth*. New York: Bloomsbury, 2015.

17 | PARADOX IN ROCKS: WANDERING FOSSILS AND TRAVELING LANDMASSES

EXOTIC TERRANES

A paradox, a paradox,
A most ingenious paradox!
We've quips and quibbles heard in flocks,
But none to beat this paradox!
A paradox, a paradox,
A most ingenious paradox.
Ha, ha, ha, ha, ha, ha, ha, ha,
This paradox.

—W. S. GILBERT, *THE PIRATES OF PENZANCE*

THE PARADOXICAL TRILOBITES

Charles Doolittle Walcott was puzzled. In the 1880s and 1890s, he had collected thousands of Cambrian fossils (mainly trilobites) in localities from California and Nevada to the Canadian Rockies to upstate New York and New England. Across most of North America, the trilobites were pretty similar in the Early Cambrian. The same primitive trilobites known as olenellids (figure 17.1) could be found from the Mojave Desert to western Newfoundland. But totally different trilobites occurred in eastern Newfoundland. Even stranger, the Early Cambrian trilobites of Scotland looked more like those of North America than those found in the rest of Great Britain or Europe. Walcott called the trilobites that ranged from the Pacific Coast to western Newfoundland the "Pacific fauna," and those found in eastern Newfoundland and Scotland the "Atlantic fauna," even

Figure 17.1 ▲

Trilobites: (*A*) typical Early Cambrian trilobite of the "Pacific fauna," *Olenellus*, one of the most primitive trilobites known; (*B*) typical Middle Cambrian trilobites of the "Pacific fauna" from the House Range of Utah. The larger trilobites are *Elrathia kingi*, while the tiny ones are the agnostid *Peronopsis interstricta*, which were apparently blind and floated in the plankton. ([*A*] Courtesy of Wikimedia Commons; [*B*] Photo by the author)

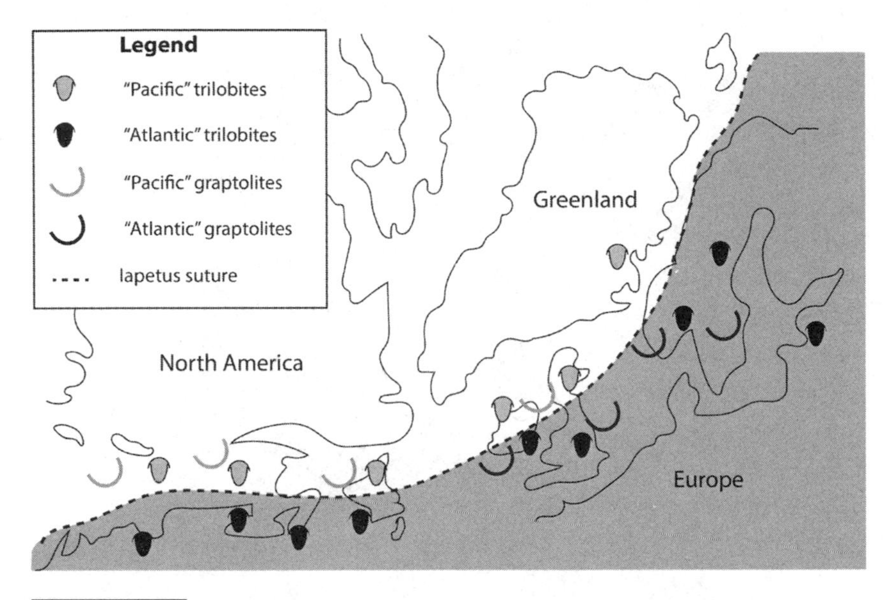

Figure 17.2 ▲

Map showing the distribution of "Atlantic fauna" and "Pacific fauna" fossils (trilobites and graptolites on each side of the modern Atlantic Ocean). The fossils of Scotland and Northern Ireland have more in common with North America ("Pacific fauna") than they do with the rest of the British Isles, while those of eastern Newfoundland, New Brunswick, Nova Scotia, and eastern Massachusetts are closer to those found in Europe than they are to those in the rest of America ("Atlantic fauna" fossils). The line separating them is the suture where the early Paleozoic predecessor of the modern Atlantic, called "Iapetus" or "proto-Atlantic," once separated the continents. (Redrawn from several sources)

though these faunal provinces didn't actually match those modern oceans, since a few of the "Pacific fauna" trilobite localities were found in western Newfoundland, which borders the modern Atlantic Ocean (figure 17.2).

It became even more puzzling in the Middle Cambrian. In the Great Basin of Utah, there were familiar forms as *Elrathia kingi* and the similar-looking *Modocia, Asaphiscus*, as well as the tiny blind trilobite *Peronopsis* (figure 17.1B). *Elrathia kingi* is the most common trilobite sold in rock shops and by commercial vendors worldwide, since it is so abundant in the Wheeler Shale of the House Range of Utah that large commercial operations used to collect them with backhoes. These same Middle Cambrian trilobites could be found from Utah to the Canadian Rockies to parts of western Newfoundland.

Yet when Walcott and other paleontologists collected from Middle Cambrian beds in eastern Massachusetts (near Braintree, home of early presidents John Adams and John Quincy Adams), New Brunswick, or eastern Newfoundland, the rocks contained trilobites completely different from those found in places as near as western New York and Pennsylvania. The fossil assemblages were dominated by the large trilobite *Paradoxides*, which reached lengths of 37 centimeters (15 inches), huge by Cambrian trilobite standards (figure 17.3). There were some trilobites in common, like the little *Peronopsis*, but the difference between the two faunal provinces was truly striking.

But what was really puzzling was that the trilobites of New Brunswick, eastern Newfoundland, and eastern Massachusetts were more similar to those of Europe than they were to those of nearby New York or Pennsylvania.

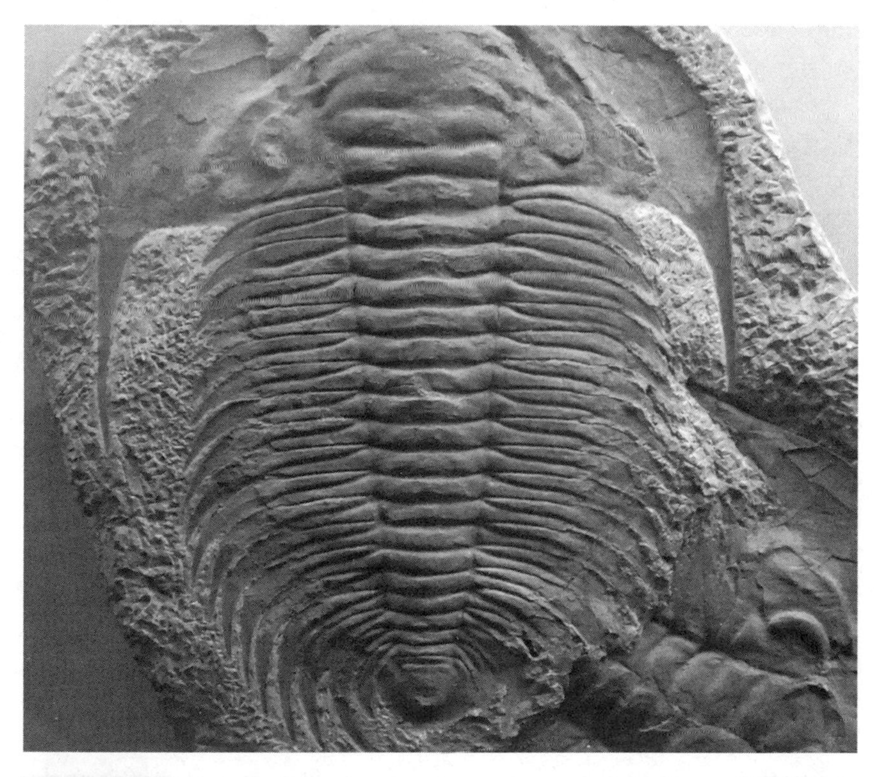

Figure 17.3 ▲
The large Middle Cambrian "Atlantic province" trilobite *Paradoxides davidis*. (Courtesy of Wikimedia Commons)

In fact, back in the 1760s Carolus Linnaeus (1707–1778), the founder of classification himself, had described a fossil from Sweden that is now called *Paradoxides paradoxissimus*. Linnaeus may not have chosen the name for the paradoxical biogeography (not discovered until much later) but for a different reason. In the 1760s, trilobites were strange fossils that fit into none of the existing groups of animals. (Another trilobite was named *Agnostus*, because the describer was agnostic about what kind of animal it was.) The spectacular giant species *Paradoxides davidis* from Manuels River Gorge on the Avalon Peninsula of Newfoundland is the same as the species recovered from the Porth y Rhaw shore of St. Davids in Pembrokeshire, Wales (hence the name *davidis*). Additional species of this genus had been found and described in France, Germany, the Czech Republic, and Poland in the early 1800s. The name *Paradoxides* was even European; it was coined by the French paleontologist Alexandre Brongniart in 1822. (He is also famous as the codiscoverer of faunal succession in France, along with Baron Georges Cuvier—see chapter 7.)

As the years went by and the collections increased, other paleontologists further demonstrated that the difference in the "Atlantic" and "Pacific" faunas was still strong in the Ordovician, and only in the Silurian were some similarities between the two marine faunas showing up. But by the Devonian, the differences had vanished completely, and the marine faunas were very similar in both Europe and North America.

What could be the reason for this paradox? Geologists and paleontologists suggested many ideas, but most assumed that there was some sort of deep but narrow ocean and a corresponding sedimentary basin that kept shallow-water trilobites from New York from reaching Massachusetts, or those from western Newfoundland from reaching eastern Newfoundland. It was thought that some time after this deep-marine trough had formed in the Cambrian, it had become narrower and narrower and eventually collapsed, so that by the Devonian it was gone, and there was only shallow water around the region. But this idea didn't make a lot of sense. Trilobites probably had planktonic larvae that could swim large distances or float in the surface currents, so why didn't they cross this narrow deep-water trough, as they could clearly float in shallow water from Utah to Quebec, or Massachusetts to Scotland? More to the point, why couldn't they cross this alleged deep-water barrier between eastern and western Newfoundland, or between New York or Massachusetts, when they could cross the width of the entire Atlantic between Scotland and western Newfoundland, or between Massachusetts and Wales?

The problem remained a puzzle for almost a century. Then, in the late 1950s the earliest pioneers of plate tectonics saw another solution. Farsighted geologists like Arthur Holmes (chapter 8) first suggested that the reason might be the motion of plates, not some peculiar marine barrier. This idea was taken up by the Canadian geologist J. Tuzo Wilson, who had already been the first to propose the idea of transform faults like the San Andreas fault (see chapter 23), or the idea that Hawaii was formed as the Pacific plate slid over and new volcanoes erupted when the plate moved on top of the hot spot.

In 1966 Wilson published a landmark paper in the leading journal *Nature*: "Did the Atlantic Close and Then Reopen?" He knew from the newly published seafloor spreading data that the modern Atlantic was a young ocean that only began to rip open and spread apart in the Late Triassic (220 Ma), early in the Age of Dinosaurs. Before this, when the continents all came together to form Pangea during the Permian and Early Triassic, the Atlantic did not exist. But the presence of different trilobite faunas suggested that there had once been a predecessor of the modern Atlantic back in the Cambrian through Devonian that vanished when all the pieces of Pangea slammed together. Wilson argued that the line that separates the Atlantic and Pacific faunas of Walcott was actually the suture between two continents that used to be separated by this extinct predecessor of the Atlantic. This long-vanished ocean was sometimes called the "proto-Atlantic" and is now called the Iapetus Ocean. When it closed in the Devonian-Permian and subsequently tore open again, it left parts of what had been on the European side of Iapetus (eastern Newfoundland, eastern New England, and a few other places) stuck to the west side of the newly formed Atlantic, and areas that had once been on the North American side (Scotland, northern Ireland) attached to Europe (figure 17.4). The opening Atlantic had ripped the crust apart roughly along the old Iapetus suture, but not exactly along the same line, leaving Scotland on one plate boundary and Massachusetts on another—and leaving eastern Newfoundland, once part of Europe, attached to western Newfoundland, which was always attached to North America. This beautifully explained Walcott's paradox.

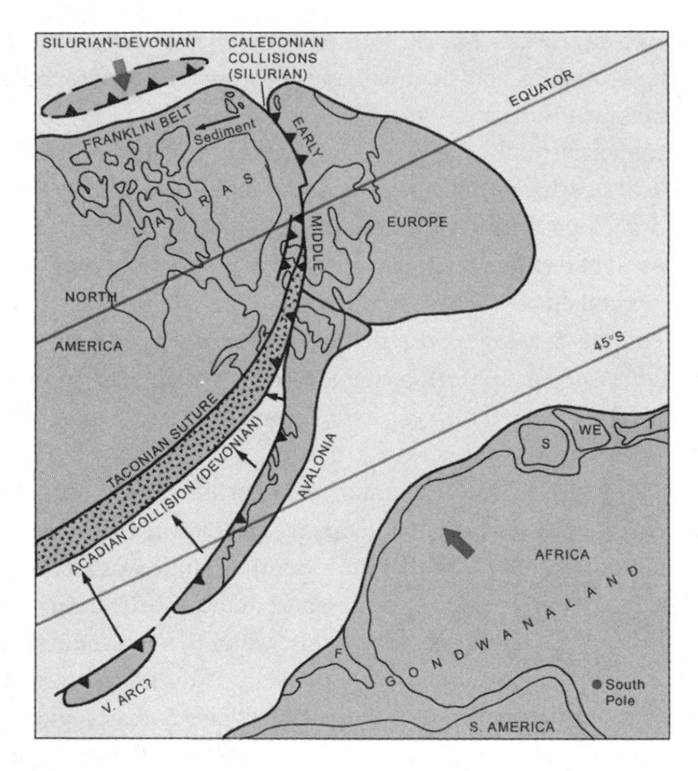

Figure 17.4 ▲
Paleogeographic map from the Silurian-Devonian showing the collision of Europe (Baltic Platform) in the Silurian Caledonian Orogeny and the Devonian collision of the Avalon Terrane along the Appalachians to form the Acadian Orogeny. (Modified from Donald R. Prothero and Robert H. Dott Jr. *Evolution of the Earth*. 8th ed. [New York: McGraw-Hill, 2010])

In his honor, this idea of closing and then reopening an ocean in roughly the same place is now called a Wilson cycle. It appears that there have been as many as five predecessors to the Atlantic Ocean, so there have been at least five Wilson cycles in that area.

THE LOST CONTINENT OF AVALONIA

In the legends of King Arthur, there was a mythical island of Avalon that played a key role in the stories. Arthur's magical sword Excalibur, which he pulled from the stone, was forged there. Arthur supposedly was taken there to recover from his wounds after fighting Mordred in the Battle of

Camlann. In some versions of the tale, Arthur also died and was buried in Avalon. The enchantress Morgana (also called Morgan le Fey) supposedly lived there as well.

The name "Avalon" comes from the Welsh "Ynys Afallon," "island of the apple," and many people have associated this mythical place with some parts of Wales, which were indeed remote to England in the time of the Arthurian legends. In Geoffrey of Monmouth's circa 1136 pseudohistory, *History of the Kings of England*, the mystical isle of Avalon is described this way:

> The island of apples which men call "The Fortunate Isle" (*Insula Pomorum quae Fortunata uocatur*) gets its name from the fact that it produces all things of itself; the fields there have no need of the ploughs of the farmers and all cultivation is lacking except what nature provides. Of its own accord it produces grain and grapes, and apple trees grow in its woods from the close-clipped grass. The ground of its own accord produces everything instead of merely grass, and people live there a hundred years or more. There nine sisters rule by a pleasing set of laws those who come to them from our country.

About 1190, a different interpretation became popular due to the work of Gerald of Wales, who argued that Arthur's Avalon was the region around Glastonbury in Somersetshire, western England. The monks of Glastonbury Abbey claimed to have found the bodies of King Arthur and his queen. According to Gerald's account,

> What is now known as Glastonbury was, in ancient times, called the Isle of Avalon. It is virtually an island, for it is completely surrounded by marshlands. In Welsh it is called *Ynys Afallach*, which means the Island of Apples and this fruit once grew in great abundance. After the Battle of Camlann, a noblewoman called Morgan, later the ruler and patroness of these parts as well as being a close blood-relation of King Arthur, carried him off to the island, now known as Glastonbury, so that his wounds could be cared for. Years ago the district had also been called *Ynys Gutrin* in Welsh, that is the Island of Glass, and from these words the invading Saxons later coined the place-name "Glastingebury."

For generations, Britons took these stories seriously, and in 1278 King Edward I (known as "Longshanks," conqueror of Wales and Scotland, and villain of Mel Gibson's William Wallace movie *Braveheart*) had the skeletons reburied at Glastonbury Abbey with high ceremony. Although this version of the Arthurian legend was convincing in the Middle Ages, today it is also regarded as pseudoarcheology. Most modern historians regard this as a publicity stunt by the monks of Glastonbury Abbey to raise funds to repair the abbey. Nevertheless, the legend motivated generations of pilgrims to visit Glastonbury Abbey, until the Reformation, when Henry VIII broke away from the Catholic Church. Many later writers connected the story of Arthur in Glastonbury with the legendary visit of Joseph of Arimathea to bring the Holy Grail to England. Others connected Avalon and Glastonbury with Earth mysteries, the puzzling ley lines, and even the myth of Atlantis. Today, there are many mythological tales and even modern romances about Avalon, such as *The Mists of Avalon, A Glastonbury Romance*, and *The Bones of Avalon*.

The Arthurian "isle of Avalon" may be only a legend, but the ancient lost continent of Avalonia is not. In recent years, the plate tectonic models for the movements of land masses on each side of the modern Atlantic have become more refined. Further studies of different fossil faunal provinces and detailed structural mapping have shown that the block that included the *Paradoxides* fauna, running from the southern Appalachians to eastern Massachusetts to New Brunswick, Nova Scotia, eastern Newfoundland, plus southern England and Wales, France, Germany, Poland, and the Czech Republic, was part of an ancient Paleozoic microcontinent now called Avalonia (figures 17.2, 17.4, and 17.5). It was not directly named for the mythical isle of Avalon, but for the Avalon Peninsula in Newfoundland. This place, in turn, was named in 1623 by Sir George Calvert when he was given a royal charter to the province of Avalon, "in imitation of Old Avalon in Somersetshire wherein Glassenbury stands, the first fruits of Christianity in Britain as the other was in that party of America." Clearly Calvert believed the myths of Arthur in Glastonbury, which were still widely held in 1623. And though no one knew it at the time the Avalon Peninsula was named, the trilobites from Manuels Gorge are indeed connected to the trilobites of Wales and the rest of England (old legendary Avalon) and the many other parts of Avalonia now scattered from eastern Europe to Canada to Georgia. Even more appropriately, one of the trilobites from Wales and western England is named *Merlinia*, after the magician of Arthurian legend.

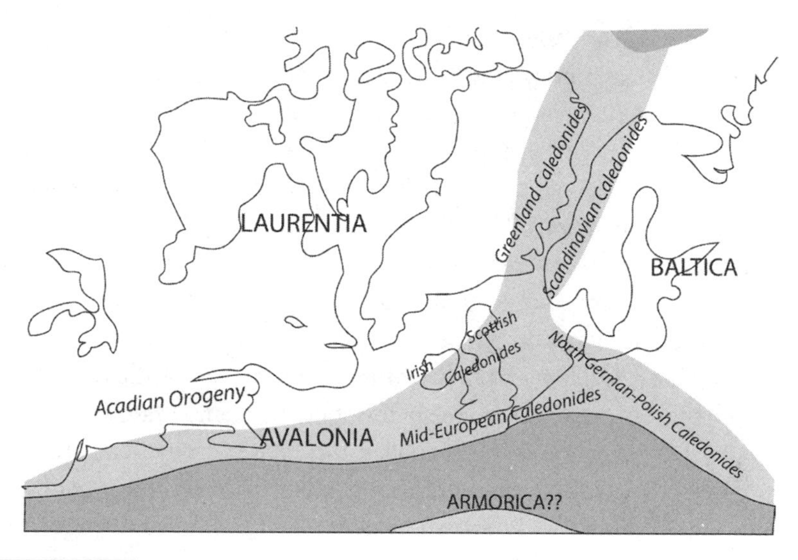

Figure 17.5 ▲
Map of the parts of the Avalon Terrane after the Caledonian and Acadian Orogenies had sutured it onto North America ("Laurentia") and Baltica. The Armorica block (today lying beneath northern Europe) later collided during the Carboniferous, forming ophiolites in Devon and Cornwall and the Carboniferous granites in those regions as well. (Redrawn from several sources)

So what happened to Avalonia? During the Cambrian and Early Ordovician, it was pulling away from Gondwana, and the trilobites of Avalonia share a lot in common with those found in the great southern supercontinent. But as it rifted away from one continent, it was rapidly approaching another: the predecessor of North America (called Laurentia). By the Late Silurian, the core of Europe (the Baltic platform, including Scandinavia and Russia) had begun to slam into Laurentia to produce a Himalayan-style collision between two continental plates and raise a mountain belt as high as the modern Himalayas. Known as the Caledonian Orogeny (after the Roman name for Scotland, *Caledonia*), its products can be seen in the crumpled and metamorphosed rocks of Scotland, the north coast of Greenland, and the Norwegian coast (figure 17.5). The river sands eroding from the Caledonian Mountains shed some of the Devonian Old Red Sandstone on top of the tilted and eroded Silurian rocks deformed by this collision, creating the famous angular unconformities at Siccar Point and Jedburgh (chapter 4).

Not long after the Caledonian collision, Avalonia began to slam into the eastern coast of Laurentia in the Devonian. It zipped shut from north to south (figure 17.5), generating another Himalayan-sized range in what is known as the Acadian Orogeny. This huge mountain range shed Devonian sandstones and shales westward across upstate New York and west-central Pennsylvania, producing the famous Catskill sedimentary sequence in the region.

In a final connection to legend, the Acadian Orogeny gets its name from the old French province of Acadia, which once included nearly all of maritime Canada, plus parts of Quebec and Maine—and the Avalon Peninsula. Those who have read American literature might recognize the name "Acadia" from the legend of Evangeline in the poem by Henry Wadsworth Longfellow. In that story, the British expelled the French-Canadians of Acadia from much of the region. Some of them ended up in southern Louisiana, where these French "Canadians" became "Cajuns"—and their cultural influence is still felt today.

EXOTIC TERRANES

The collision of the Avalon Terrane with Laurentia is a well-documented example of what are known as exotic terranes, blocks of crustal material that have clearly come from someplace else. The entire Appalachian region is built of chunks of other continents that came to Laurentia at different times. The first major collision was a block of crustal material known as the Piedmont Terrane, which can be found in parts of eastern Canada, in the Hudson Valley of New York, and down to the Piedmont of the Appalachian foothills in Virginia, Georgia, and the Carolinas (see figure 17.2). It arrived in the Late Ordovician in an event known as the Taconic Orogeny, which uplifted and deformed the rocks of the modern Taconic Mountains in New York.

Then came the collision of Baltica to form the Caledonian Orogeny in the Late Silurian (influencing the northern coast of Laurentia), followed by the Devonian Acadian Orogeny that added the Avalon Terrane from maritime Canada through eastern Massachusetts and down to the Carolina Slate Belt in the South. The final event was the collision of the African portion of Gondwana with the East Coast to crumple up the Appalachians in the Late Carboniferous (Pennsylvanian). This event also closed the predecessors of the Atlantic for good, and joined Laurentia or North America to the supercontinent of Pangea. The Appalachians were formed in this huge Himalayan-style collision over 300 millions of years ago and have been slowly eroding ever since.

If it is not staggering enough to imagine eastern Massachusetts or the eastern Carolinas as pieces of other continents, even more amazing is the assembly of exotic terranes that make up the Pacific Coast of North America. All of Alaska and most of British Columbia, Washington, Oregon, Idaho, Nevada, and California are exotic terranes (figure 17.6). Once again, the first evidence for this came from fossils. For decades, geologists and paleontologists had been puzzled about some of the fossils that had been collected on the Pacific Coast, from corals found in northern California and British Columbia to the rice grain–shaped, single-celled amoebas known as fusulinid foraminiferans, found all the way from Alaska to northern California. When these fossils were examined closely, it was clear that they came from the tropics of the Southern Hemisphere—and from the other side of the Pacific Ocean in what was then the Tethys Seaway that ran from Indonesia to Gibraltar!

At first geologists and paleontologists refused to accept the idea that huge blocks of crust had sailed across the entire Pacific in the past 250 million years, all the way from Indonesia to British Columbia, but now the evidence is clear. It is supported not only by the fossils, but also by the tectonic evidence of how Alaska, British Columbia, and most of the Pacific Coast states of are assembled as tectonic blocks, all slapped together along major faults. Finally, it was confirmed by paleomagnetic data that showed these blocks had indeed come from south of the equator and apparently from as far away as the modern location of Indonesia as well.

Today, we can reconstruct where all these pieces of the puzzle came from and compare that with where they are now (figure 17.7). One of the first to arrive in the Devonian-Mississippian Antler Orogeny was a block (the Antler Terrane) that runs across central Nevada and includes rocks found in the western foothills of the Sierra Nevada and the eastern side of the Klamath Mountains near Mount Shasta. The next big block to arrive was the Sonomia Terrane in the Permo-Triassic Sonoma Orogeny. It makes up the entire northwest corner of Nevada plus additional blocks in the Sierra foothills and Klamaths that were slammed underneath the blocks that arrived in the Antler Orogeny. Not far behind it were many more terranes, including Stikinia and Quesnellia, which make up the core of British Columbia and extend into Alaska, and Wrangellia, which makes up the Wrangel Range in

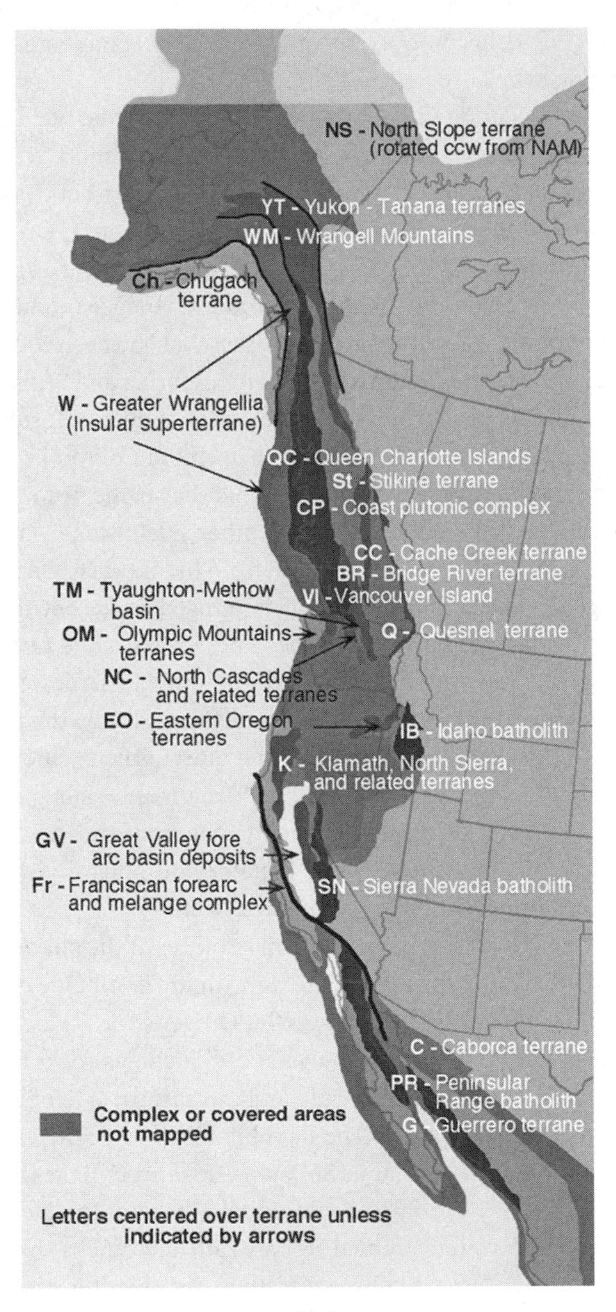

Figure 17.6 ▲

Map of exotic terranes in the Pacific rim of North America (Redrawn from several sources)

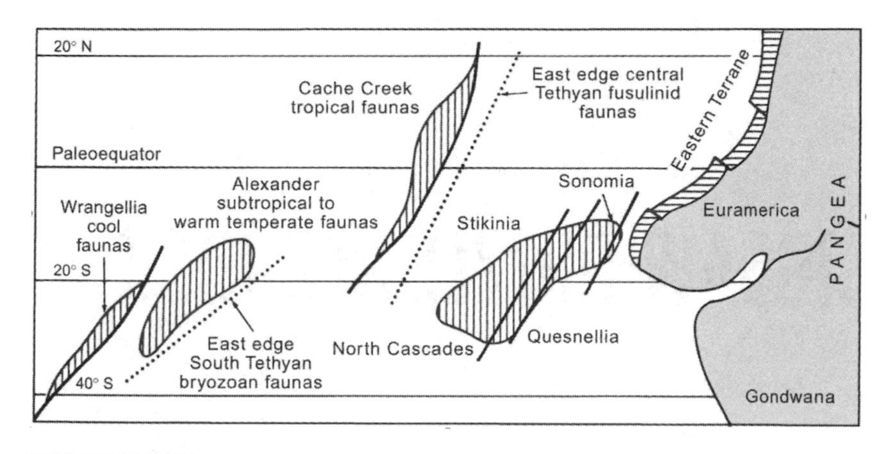

Figure 17.7 ▲

Map showing the original location and distribution of exotic terranes in the proto-Pacific, all of which are now collided and crumpled into Alaska, British Columbia, and coastal Washington, Oregon, and California. During the Permian, many of them, such as the Wrangellia and Alexander Terranes, began in what was then the southwestern Pacific, based on the distinctive fossils from the Tethys Seaway that they carry. (Modified from Donald R. Prothero and Robert H. Dott Jr., *Evolution of the Earth.*, 8th ed. [New York: McGraw-Hill, 2010])

Alaska (including America's highest mountain, Denali or Mount McKinley) and northwestern British Columbia and the Alaska Panhandle. Most of these blocks arrived in the Age of Dinosaurs and were in place by the Cretaceous—carrying their Permian fossils from the tropical southwestern Pacific with them.

So the next time you are in Alaska, British Columbia, Washington, Oregon, Nevada, or California, just remember: you're not in North America. You're actually in Fiji or Indonesia.

FOR FURTHER READING

Fortey, Richard. *Trilobites: Eyewitness to Evolution*. New York: Vintage, 2001.

Levi-Setti, Riccardo. *Trilobites: A Visual Journey*. Chicago: University of Chicago Press, 2014.

Prothero, Donald R. *Bringing Fossils to Life: An Introduction to Paleobiology*. 3rd ed. New York: Columbia University Press, 2013.

Prothero, Donald R., and Robert H. Dott Jr. *Evolution of the Earth*. 8th ed. New York: McGraw-Hill, 2010.

18 ALFRED WEGENER AND CONTINENTAL DRIFT

JIGSAW-PUZZLE BEDROCK

A new theory is guilty until proven innocent, and the pre-existing theory innocent until proven guilty . . . Continental drift was guilty until proven innocent.

—DAVID M. RAUP

HE WAS DESPISED AND REJECTED . . .

Christmas 1910. A 30-year-old German meteorologist named Alfred Wegener (figure 18.1) happened to glance through a world atlas that a friend had received as a Christmas gift. Flipping through the pages, he saw the amazing match between the coastlines of South America and Africa. It was like a light bulb going off in his head. Why did these two continents, separated by the huge South Atlantic Ocean, seem to fit so well? In fact, Wegener was not the first to notice this match. As early as the 1500s people had commented on it, almost as soon as the first decent maps of the Atlantic were available.

In Wegener's own words:

> The first concept of continental drift first came to me as far back as 1910, when considering the map of the world, under the direct impression produced by the congruence of the coast lines on either side of the Atlantic. At first I did not pay attention to the ideas because I regarded it as improbable. In the fall

Figure 18.1 ▲
Alfred Wegener on his fourth and last Arctic expedition in 1930. (Courtesy of Wikimedia Commons)

> of 1911, I came quite accidentally upon a synoptic report in which I learned for the first time of palaeontological evidence for a former land bridge between Brazil and Africa. As a result I undertook a cursory examination of relevant research in the fields of geology and palaeontology, and this provided immediately such weighty corroboration that a conviction of the fundamental soundness of the idea took root in my mind.

But Wegener did not casually notice the pattern and then move on to other things, as everyone else had done for 400 years. He was building a career as a budding scientist, with many years experience in meteorology and climate. That same year he wrote the standard German textbook in the science, *Thermodynamics of the Atmosphere*, so he clearly knew his stuff. Nor was he green and inexperienced. At the young age of 26, he had organized and led the first of four expeditions to Greenland to study polar climate and weather.

Despite his heavy time commitment to meteorological and polar research and teaching classes at the University of Marburg, Wegener continued to look for evidence that the continents once fit together. In 1908 he began to work with the legendary climatologist Wladimir Köppen, who created the standard classification of climate regions and zones that is still used today. (I teach it every year in my meteorology class.) Wegener and Köppen searched the scientific literature and found that the climate zones of the Permian Period (250–300 Ma) don't make any sense as they are distributed on the continents today. They began to compile evidence that these climatically sensitive rocks suggest that the continents had moved. By 1912 Wegener had given a few lectures on his evidence for continental drift, then he published three short papers on the evidence in German geographical journals. In 1913 Wegener married Köppen's daughter—and then led a second expedition to Greenland, where he spent the winter on the ice and almost died there before he and his companion were rescued.

On June 28, 1914, Archduke Franz Ferdinand was assassinated, and soon World War I broke out. Like every other able-bodied man in Germany at the time, Wegener was called to serve in the Kaiser's army. He was wounded twice (once in the neck) before the German High Command decided that as a trained meteorologist, he was more useful to them in their weather stations than as trench fodder. As he traveled from one German weather station to another, he continued to document his ideas. His book *On the Origin of Continents and Oceans* was published late in 1915, but few people read it or even saw it during the wartime restrictions. In it, he published the first-ever maps (figure 18.2) to show how the continents had drifted apart since the Permian (since 250 Ma) and their configuration in the supercontinent of Pangea, with the southern half known as Gondwana and the northern part called Laurasia (Laurentia plus Eurasia). Despite his duties, Wegener found that being an army meteorologist was a good posting; he had published another 20 papers in meteorology and climatology by the war's end.

Once the war was over, Wegener took several different jobs in Hamburg, then he accepted a permanent post at the University of Graz. He used this time to write a book with Köppen about climates in the geological past, based on the evidence he had originally collected to support his ideas about continental drift. Still, Wegener was mostly unknown outside Germany, especially because his works did not get translated into English until 1925.

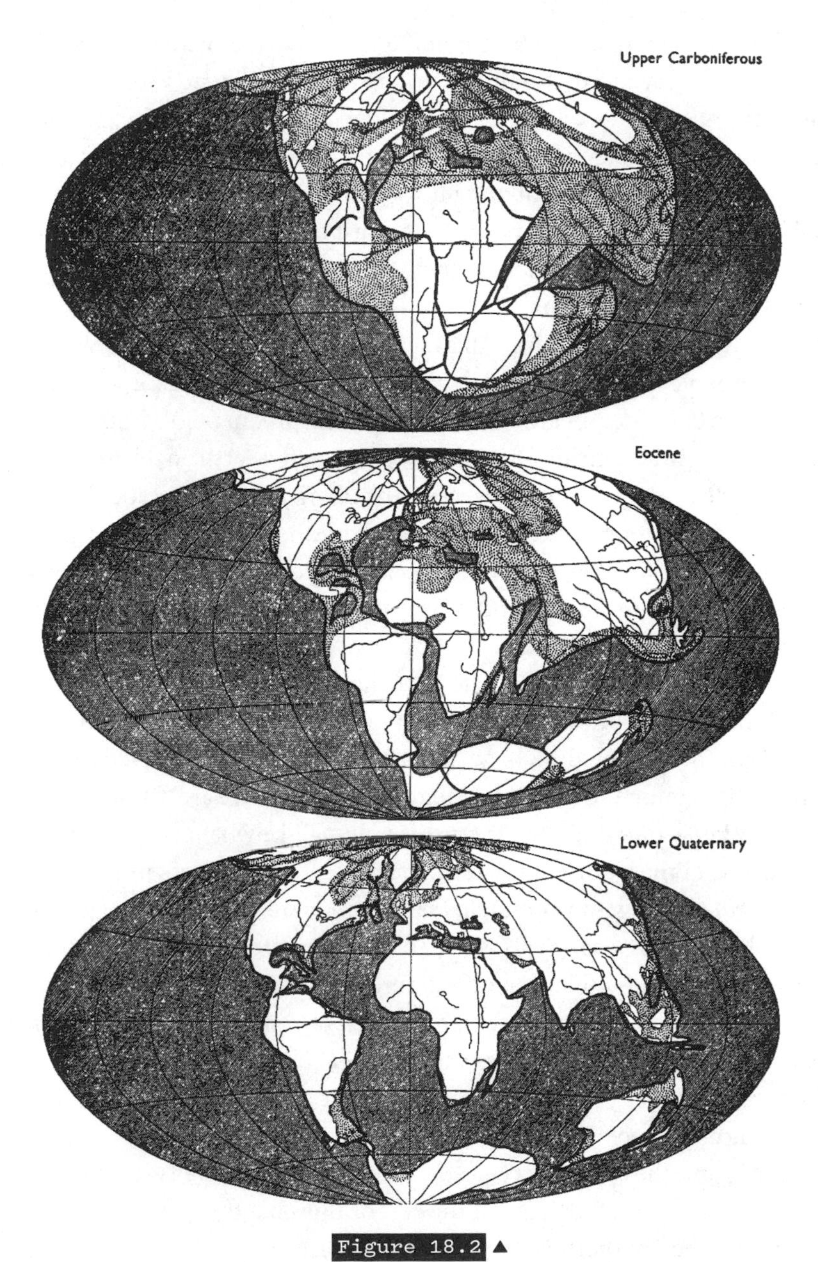

Figure 18.2 ▲
Wegener's 1915 map of the movement of continents. (From Wegener 1915)

After more than 14 years of being ignored by mainstream geologists, Wegener was invited to speak about his ideas at the 1926 meeting of the American Association of Petroleum Geologists in New York City. This symposium had been organized by his opponents as a chance to ridicule his ideas, and Wegener was walking into a lion's den. Only the chairman who invited him gave him a fair hearing; the rest of the audience was scornful and dismissive. His ideas sounded crackpot to them, based on what they knew of geology at the time.

Why did they not take him or his ideas seriously? For one thing, Wegener was not a geologist but a meteorologist and climatologist. There is some justification in being skeptical of outsiders dabbling in your field without having a full training in it. I run into crackpot ideas about the earth on the Internet all the time. They range from flat-earthers to hollow-earthers to geocentrists to young-earth creationists to people who believe the earth is expanding. Anyone with a basic first few courses in geology can easily see why these people are wrong, and it is especially obvious to those geologists who have real field experience and are not just dreaming up ideas based on secondhand sources.

In addition, Wegener's ideas did have their flaws. He argued that the continents had moved around the globe, but geologists argued that if Wegener were right, there should be huge areas of oceanic crust crumpled up like a carpet when continents plowed through them—and there are few such places. (We now know that the oceanic crust is nothing like what people thought at that time and that it usually slides beneath other continents in subduction zones, rather than getting crumpled up into mountain belts.) Wegener could not explain how the continents moved or what powered their motion, and the ideas he did propose (like centrifugal force) were geophysically impossible. Also, the rates of plate motion were too high (250 centimeters per year), when today we know that most plates move only about 1 percent as fast as that (2.5 centimeters per year). To be fair, when he proposed that rate around 1915, the dating of the geological time scale was just beginning, so no one knew how far back in time the Permian occurred, which was when Pangea was assembled.

Finally, as we shall discuss in the rest of this chapter, the best evidence came from the Southern Hemisphere. Yet nearly all geologists at that time were living and working in North America and Europe, with only a handful coming from less-developed countries around the world. In the early twentieth century, it was very slow and expensive to take an ocean liner down to Brazil or South Africa, so very few geologists had actually been to these regions and seen the rocks for themselves. Most geologists were limited to reading descriptions in journals and books and reviewing a handful of muddy black-and-white photographs that did not do justice to the brilliant colors and striking similarity of rocks in the Southern Hemisphere. Geologists who saw the rocks firsthand, like South African Alexander du Toit, were the biggest supporters of continental drift, but they were also outsiders whose ideas were seldom presented in the geological meetings of North American or European geologists. One of the few European supporters of the idea, Arthur Holmes (chapter 8), had worked in Africa, so he also knew the rocks from personal experience. Thus, the entire idea of continental drift remained a crackpot notion for another 30 to 40 years.

Meanwhile, Wegener did not sit in a corner and sulk over the poor reception of his grand ideas. He continued his meteorological research as a polar explorer, and in 1929 he led his third expedition to Greenland. The next year he led his fourth and largest expedition—and also his last. It was equipped with a wide range of weather equipment, plus a propeller-driven snow sled and other devices. There was one remote station in the middle of the Greenland ice sheet known as *Eismitte* ("mid-ice" in German), one of the coldest locations in the Northern Hemisphere. It has an average mean temperature of −30°C (−22°F) and routinely reaches temperatures of −62°C (−80°F) in winter. Being near the Arctic Circle, it does not see the sun from November 23 to January 20. Station Eismitte was so remote that a trip for supplies was both dangerous and lengthy.

In November 1930, Wegener and his partner Rasmus Villumsen were on their way back from a supply run when they got caught in horrific blizzards and extremely cold temperatures. Wegener died, possibly from a heart attack (he was a heavy smoker) or possibly because hypothermia overcame him and he froze to death. Villumsen buried him in the snow with skis to mark his grave and then was never seen again. A team later found Wegener's grave and reburied it with a cross to mark it. There his body still lies under 100 meters (330 feet) of ice, moving along with the flowing ice of the Greenland ice sheet. Wegener died just past his fiftieth birthday, unnoticed and unmourned by the geological community. If he had lived another 30 years, he might have witnessed the vindication of his ideas—but luck and fate were not with him. Instead, he is yet another example of a genius with a great idea who died scorned and unappreciated, never living to see his work make it from crackpot to scientific paradigm.

PUZZLE #1: JIGSAW-PUZZLE ROCKS

What was this evidence that convinced Wegener and so many geologists in the Southern Hemisphere? There were two main areas of evidence: the rocks of the Permian Period and the ancient Precambrian basement rocks beneath most of the southern continents.

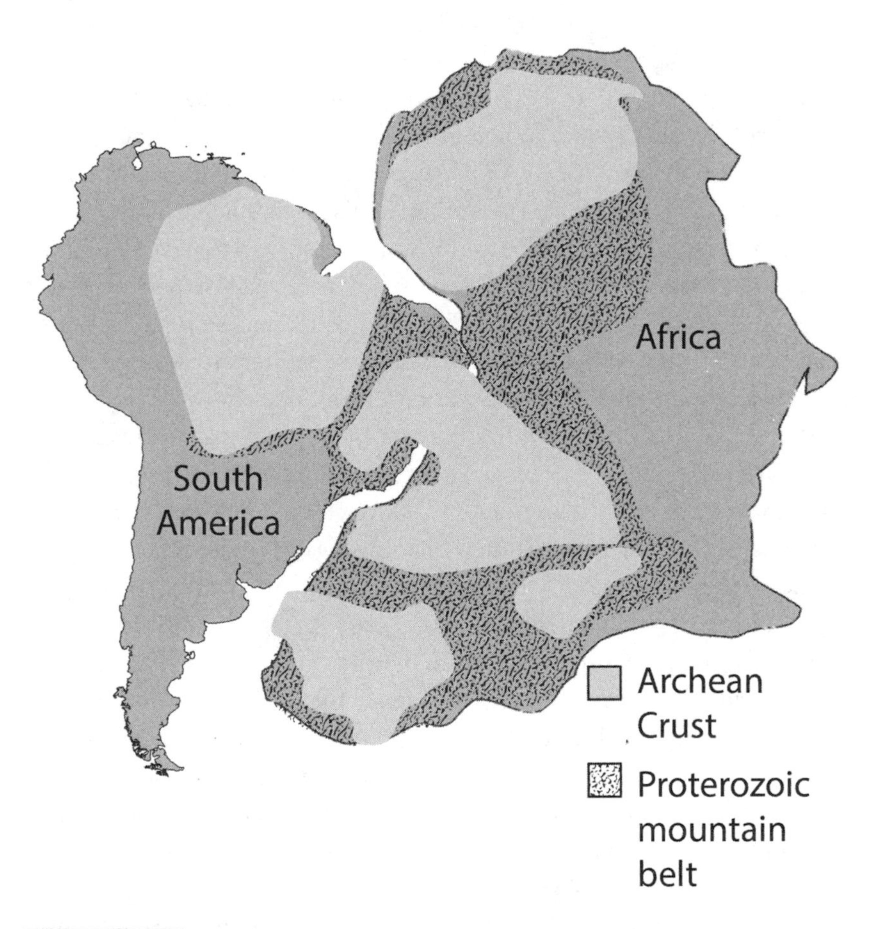

Figure 18.3 ▲

The jigsaw-puzzle match of the Precambrian bedrock of Africa and South America. Both continents began as small protocontinents made of Archean crust (older than 2.5 billion years old) that slammed together to form Proterozoic mountain belts of the rocks trapped between them. Today, these features are ripped apart, so the Archean cores and the Proterozoic mountain belts are found on each side of the Atlantic, and their patterns fit together like pieces of a jigsaw puzzle. (Redrawn from several sources)

One of the most impressive pieces of evidence is the ancient bedrock of the continents that made up the Gondwana supercontinent, namely South America, Africa, Australia, India, and Antarctica (figure 18.3). Strip away the covering of vegetation and younger rocks, and you get down to the basement rock that underlies the continent. It is made of ancient protocontinents from the Archean. They are squashed together, and in between they trap Proterozoic mountain belt rocks that formed on their edges before the protocontinents slammed together. Every continent has some combination of these Archean protocontinental core rocks, amalgamated into a larger continent with the Proterozoic rocks crushed between them.

What is striking about the basement rocks of South America and Africa, however, is that these ancient bedrock patterns are chopped up abruptly by the Atlantic Ocean. Once you put South America up against Africa, however, the Archean cores fit together, as do the Proterozoic belts caught between them, like pieces of a jigsaw puzzle.

There is simply no way to explain away this amazing match. Yet for 40 years, geologists tried to do so, either by claiming that it was a coincidence or that the bedrock was not as similar as was claimed or simply by ignoring it.

PUZZLE #2: CLIMATE BELTS IN THE WRONG PLACE

If you travel to South Africa or Brazil or Antarctica or India or Australia, you will see a strikingly similar sequence of rocks. They include a distinctive Carboniferous sandstone sequence with coal beds, followed by the Lower-Middle Permian glacial till, overlain by a thick sequence of Permian-Triassic red beds full of fossils of reptiles and protomammals, and finally huge Jurassic lava flows capping the entire sequence. I've heard several geologists remark that if you didn't read the formation names or hear the local geologists speaking either Afrikaans or Portuguese, you could not tell whether you were in South Africa or Brazil.

Even more revealing is where these distinctive deposits are found (figure 18.4). The coal deposits of the Permian, which should be found only in the equatorial rain forest belt, are far outside that region today. Likewise, the Permian desert dune deposits are not located in the subtropical high-pressure belts between 10° and 40° north and south of the equator, like they are in the modern world's deserts. But if you put the continents back into the Pangea configuration, rather than the modern

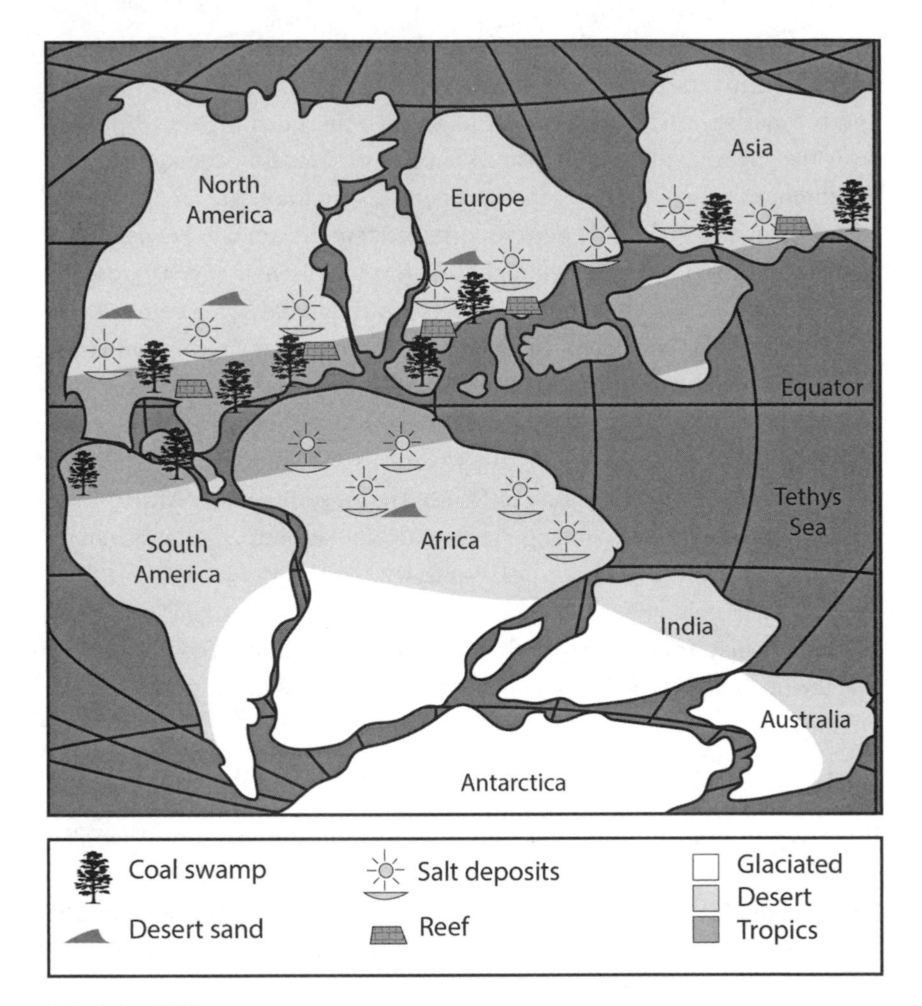

Figure 18.4 ▲

The climatic belts of the Permian (glaciers on the poles, coal swamps in the tropical rain forests, dune sands in the subtropical high-pressure belts) only makes sense in the Pangea configuration of the continents. These same deposits are completely out of place on a map of their modern locations. (Redrawn from several sources)

position of the continents, then all the Permian coal beds are in the tropical rain forest belt, and all the Permian dune sands are in the subtropical high-pressure belt, where they belong.

Most impressive of all are thick glacial beds found on many of the Gondwana continents, such as the Dwyka tillite of South Africa (figures 18.5

Figure 18.5 ▲

Outcrop of the Permian Dwyka tillite in South Africa, formed by Permian glaciers on Gondwana. Like most other glacial deposits, these tills are formed of large cobbles and boulders mixed with fine sands and muds, with no sorting or bedding typical of water-laid deposits. (Courtesy of Wikimedia Commons)

and 18.6) and equivalents on other continents, including South America, the Indian subcontinent, Australia, and Antarctica. These only make sense in light of the Permian Gondwana location of these land masses. If you tried to plot the distribution of those ice sheets on the modern continents, then you'd have a Permian ice sheet that straddles the South Atlantic and most of the Indian Ocean and reaches across the equator to some parts of India. This clearly makes no sense paleoclimatically.

Even more remarkable were the scratches and gouges created as the glaciers dragged huge rocks across the landscape (see figure 18.6). In some cases, the gouges originate in South Africa and then line up with similar scratches in South America. For this to make sense within the context of the modern globe, the Permian glaciers would have had to jump into the Atlantic and cross it in a straight line, then jump right back onto land in that same straight course. This is absurd as well. Only if the continents were in their Gondwana configuration, with no Atlantic Ocean to cross, would these aligned scratches make any sense.

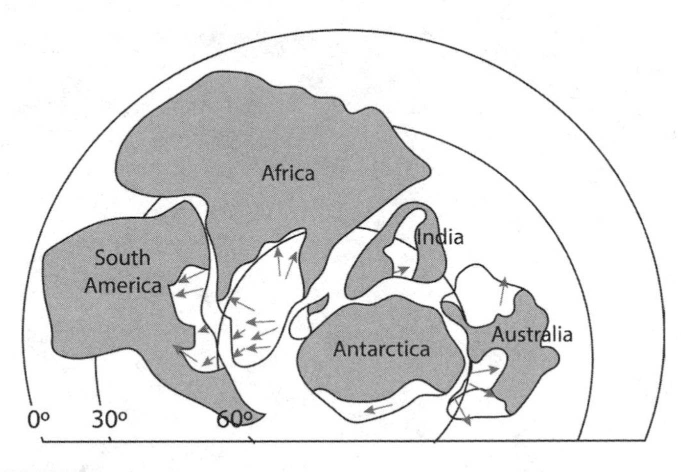

Figure 18.6 ▲
Map showing glaciated areas of Gondwana during the Permian (white), and how the glacial scratches line up between southwestern Africa and Brazil. (Redrawn from several sources)

In addition to all this evidence from late Paleozoic-early Mesozoic Gondwana rocks, the fossils are even more revealing. Nearly every Permian deposit in Gondwana contains leaves of the primitive extinct seed fern, *Glossopteris*—including Antarctica, Australia, and Madagascar (figure 18.7). Then there is the little half-meter-long reptile *Mesosaurus*, only known from lake beds in South Africa and Brazil—and too small to swim across the modern South Atlantic. Almost every Gondwana continent yields fossils of the little beaked protomammal (formerly but incorrectly called a "mammal-like reptile") known as *Lystrosaurus*. It was already known from Africa, South America, and India, and its discovery in Antarctica in 1969 was considered by some to be the clinching evidence for continental drift. Finally, there is the bear-sized predatory protomammal *Cynognathus*, which occurs not only in South Africa and South America, but also up in Russia in the Late Permian.

All of this evidence convinced geologists like Du Toit and Holmes, but skeptical Northern Hemisphere geologists tried to dismiss it or explain it away. The peculiar distribution of Permian Gondwana plants and animals was attributed to land bridges or animals rafting across oceans. The matches in the rocks themselves were either ignored (because they were not that convincing in tiny black-and-white journal photos) or dismissed as circumstantial. In short, almost none of this evidence, which seems

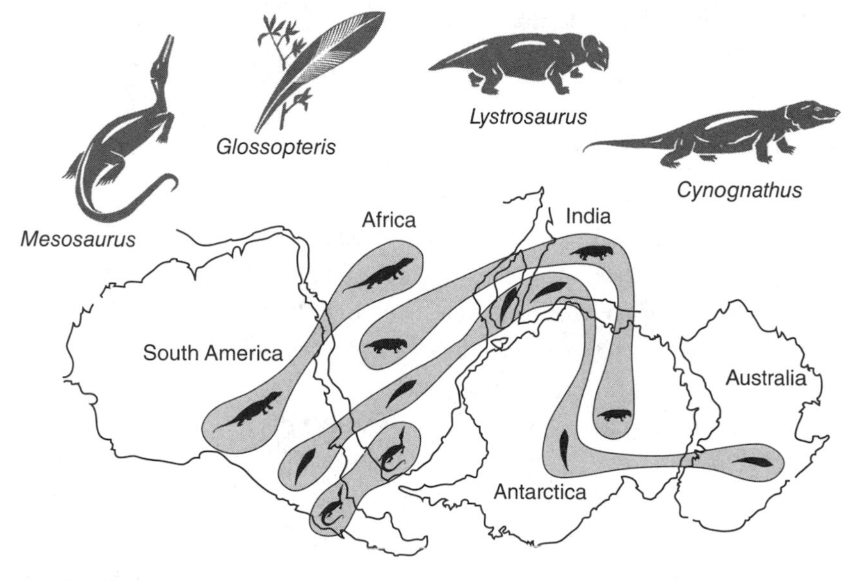

Figure 18.7 ▲
Map showing the Gondwana distribution of the Permian seed fern *Glossopteris*, the little lake-dwelling reptile *Mesosaurus*, the herbivorous protomammal *Lystrosaurus*, and the large predatory protomammal *Cynognathus*. (Redrawn from several sources)

overwhelming to geologists today, was given a fair hearing by most of the geologists from the time of the publication of Wegener's book in 1915 until the early 1960s—almost 50 years of appalling blindness by the majority of the world's geologists.

Continental drift theory was considered fully crackpot by the most powerful American and European geologists of the late 1940s and early 1950s. The famous biologist and documentary filmmaker David Attenborough, who went to university in the late 1940s, recalled: "I once asked one of my lecturers why he was not talking to us about continental drift and I was told, sneeringly, that if I could prove there was a force that could move continents, then he might think about it. The idea was moonshine, I was informed." In 1949 the American Museum of Natural History held a symposium attacking the evidence by talking about land bridges and dismissing the similarity between rocks on different continents (even though most geologists who did so had never seen these rocks firsthand). The symposium was not published until 3 years later, but in retrospect, it is an amazing monument to being shortsighted and wrong just as the world was about to change beneath our feet.

SOLUTION IN THE OCEAN DEEPS

Meanwhile, evidence was coming from an entirely new direction: the bottom of the sea. Remember how geologists dismissed Wegener's arguments by saying that ocean floor rocks should be pushed in front of the drifting continents like the snow in front of a snowplow? It turns out that their argument was completely wrong, because no one really knew anything about the crust of the earth beneath the oceans.

In fact, almost nothing was known about the deep ocean before World War II. Oceans make up almost 71 percent of the earth's surface, yet we knew very little about them until this postwar period. Then the importance of submarine warfare made it apparent to the world's navies that we needed to really understand the deep ocean. After the war ended, most of the countries cut back on military spending, but the United States and a few other countries began pouring funds into oceanographic institutions to correct this long-term blind spot in human knowledge. Not only funding, but also war-surplus navy vessels came to these institutions to be repurposed as oceanographic vessels (rather than being cut up for scrap).

By the late 1940s and early 1950s, the major oceanographic institutions (Scripps in San Diego, Woods Hole in Massachusetts, Lamont-Doherty in New York, and a few others) all had vessels sailing around the world all year, collecting data on the temperature, salinity, density, and chemistry of ocean water; on the depth to the seafloor and the nature of the rocks and sediments beneath the seafloor; dropping long tubes called piston corers off the side of the boat to take 33-foot-long cores of sediment recording millions of years of oceanic history; and towing torpedo-like proton-precession magnetometers (once used to hunt submarines) to measure the magnetism of the ocean floor rocks.

Over the course of the 1950s and 1960s, oceanographic research managed to solve a lot of the mysteries of the world's largest bodies of water. Marie Tharp and Bruce Heezen (see figure 22.1) at Lamont-Doherty Geological Observatory (now Lamont-Doherty Earth Observatory) completed the first map of the ocean floor. In the process, Tharp documented that there was a giant rift valley down the middle of the mid-ocean ridge, which proved to her that the plates were spreading apart (see chapter 21). Detailed research by many vessels showed not only how deep the ocean basin is around the world, but also what lies beneath the surface of the ocean floor. Deep-sea sediment cores showed how the oceans had changed through time and how climate had changed around the world, even unlocking the causes of the Ice Ages (chapter 25). But most important of all, the magnetometers towed behind the ships showed a peculiar pattern of magnetized rock that eventually proved the reality of seafloor spreading in 1963 (chapter 21). After that discovery, the plate tectonic revolution was in high gear and permanently transformed the entire geosciences.

Wegener never lived to see that happen a full 33 years after he died, but many of his later critics did live to see it. Some of them (especially old-guard oil geologists) refused to accept plate tectonics and eventually died off or stopped doing research. Others (like the authors of the infamous 1949 American Museum symposium) grudgingly accepted that they had been wrong. A few of them embraced the new model—the legendary Columbia University stratigrapher Marshall Kay, who found that his entire life's work had become obsolete, cheerfully started to revisit his work in light of plate tectonics, even though he was in his sixties. Finally Wegener received honor and praise for being so far ahead of his time, a prophet who proved to be right at the end.

FOR FURTHER READING

Greene, Mott T. *Alfred Wegener: Science, Exploration, and the Theory of Continental Drift*. Baltimore: Johns Hopkins University Press, 2015.

McCoy, Roger M. *Ending in Ice: The Revolutionary Ideas and Tragic Expedition of Alfred Wegener*. Chicago: University of Chicago Press, 2006.

Oreskes, Naomi. *Plate Tectonics: An Insider's History of the Modern Theory of the Earth*. New York: Westbury, 2003.

——. *The Rejection of Continental Drift: Theory and Method in American Science*. New York: Oxford University Press, 1999.

Wegener, Alfred. *The Origin of Continents and Oceans*. New York: Dover, 2011.

19 THE CRETACEOUS SEAWAY AND THE GREENHOUSE PLANET

CHALK

There'll be blue birds over the White Cliffs of Dover
Tomorrow, just you wait and see.
There'll be love and laughter and peace ever after
Tomorrow, when the world is free . . .

—NAT BURTON, "THE WHITE CLIFFS OF DOVER"

THE WHITE CLIFFS OF DOVER

Crossing the English Channel from eastern France or Belgium to southeastern England, the first thing you see on the horizon is the legendary White Cliffs (figure 19.1). They have symbolized the notion of England as an island-fortress, and psychologically they serve as castle walls or ramparts against invaders from the continent. However imposing they may seem, invaders have always found a way around them. The armies of William the Conqueror found a low spot at Pevensey between the White Cliffs to land their boats, and then moved inland to conquer England after the Battle of Hastings in 1066. Nevertheless, the White Cliffs were a welcome sight to British troops fleeing the beaches of Dunkirk in 1940, or to Allied bomber crews trying to nurse their crippled aircraft home. During the Battle of Britain, they served as a strategic high point for observers reporting waves of German aircraft on the way, and also for the location of the secret radar towers that allowed the British to know of the incoming aircraft well in advance. The White Cliffs were also a sentimental symbol

Figure 19.1 ▲

The White Cliffs of Dover: (*A*) Viewed from the Dover Strait; (*B*) the overlook at its highest point, Beachy Head. (Courtesy of Wikimedia Commons) of England, as typified by the old World War II ballad, "The White Cliffs of Dover," which nostalgically transported homesick British troops back from the battlefield in Europe to a peaceful life in England. In 2005 a poll of listeners in *Radio Times* voted them the third greatest natural wonder in Britain. One of the old Roman names for England, Albion, comes from the Latin word "white" and probably refers to the chalk cliffs that the Roman invaders had to skirt.

WHAT IS CHALK?

The White Cliffs get their color from their bedrock, which is known as chalk. When most people hear the word "chalk," they think of the sticks of white powdery material used to write on blackboards for generations. Even though real chalk was once used for that purpose, most modern blackboard "chalk" is not made of chalk at all, but powdered gypsum compressed into sticks.

Real chalk is soft, white, porous limestone composed of the mineral calcite (calcium carbonate, or $CaCO_3$). It forms under deep-marine conditions from the gradual accumulation of minute calcite shells (coccoliths) shed from micro-algae known as coccolithophores (figure 19.2). Flint (a type of chert typical of chalk) is very common as bands parallel to the bedding or as nodules embedded in chalk (figure 19.3). It was probably derived from sponge spicules or other siliceous organisms as water was expelled upward during compaction. Flint is often deposited around larger fossils that may be silicified (i.e., calcite replaced molecule by molecule by silica).

The Chalk is rich in fossils that can be collected on the beaches in many places. It is famous for its heart urchins and sea urchins, intensively studied and written about by paleontologists, and yields a variety of clams and oysters, especially the weirdly coiled oysters *Gryphaea* and *Exogyra*, which also have been widely studied by paleontologists. The formation also yields a few poorly preserved ammonites, as well as shark teeth and various fish fossils.

Chalk has greater resistance to weathering and slumping than the clays with which it is usually associated, thus forming tall steep cliffs where chalk ridges meet the sea. Chalk hills, known as "chalk downland," usually form where bands of chalk reach the surface at an angle, so it forms

Figure 19.2 ▲

Chalk is made of tiny calcareous algae known as coccolithophorids that secrete coccoliths, button-shaped plates that surround them and measure only a few microns across. (Courtesy of Wikimedia Commons)

Figure 19.3 ▲

Black flint layers are common in the Chalk. (Photo by the author) a scarp slope. Because chalk is well fractured, it can hold a large volume of groundwater, providing a natural reservoir that releases water slowly through dry seasons.

The chalk outcrops are not restricted to southeastern England. The belt of chalk actually extends to the Alabaster Coast of Normandy and Cape Blanc Nez across the Dover Strait. There are chalk beds beneath the Champagne region of France that produces the soils for planting Champagne grapes and natural caverns for wine storage. The Chalk extends farther east through Belgium and countries to the northeast, including Jasmund National Park in Germany and Mons Klint in Denmark. The Chalk was such a distinctive and widespread unit across so much of Europe that in 1822, Jean-Baptiste-Julien D'Omalius D'Halloy named the Cretaceous Period after it, from the Latin word *creta*, which means "chalk."

THE SHALLOW SEAS OF THE CRETACEOUS GREENHOUSE WORLD

Ninety million years ago what is now the chalk downland of northern Europe was calcareous ooze accumulating at the bottom of a great sea. Chalk was one of the earliest rocks made up of submicroscopic particles to be studied under the electron microscope, when it was found to be composed almost entirely of coccoliths. Their shells were made of calcite extracted from the rich seawater. As they died, a substantial layer gradually built up over millions of years and, through the weight of overlying sediments, eventually became consolidated into rock. Later movements of the earth's crust related to the formation of the Alps raised these former seafloor deposits above sea level.

The Chalk is also a product of a much bigger phenomenon: the global greenhouse world of the Late Cretaceous. During the latter half of the Age of Dinosaurs, the climate was so warm that dinosaurs roamed above the Arctic and Antarctic circles. There was virtually no ice and snow anywhere at that time, since the atmospheric carbon dioxide levels were about 2,000 parts per million (today the level is over 400 parts per million and climbing). The melting of all those icecaps resulted in extremely high sea levels, and shallow seas drowned many of the continents. The shallow chalk-rich seas of the Cretaceous drowned most of Europe, as the distribution of the Chalk not only in England but in Belgium and France attests. The rising sea level flooded the Central Plains of North America, producing the "Western Interior Cretaceous Seaway," which connected the Arctic Ocean with the Gulf of Mexico (figure 19.4). Those seas were filled with ammonites, clams, snails, and huge fish and marine reptiles, which are found today in chalk beds such as the Niobrara Chalk in western Kansas (figure 19.5) or the Austin Chalk of central Texas.

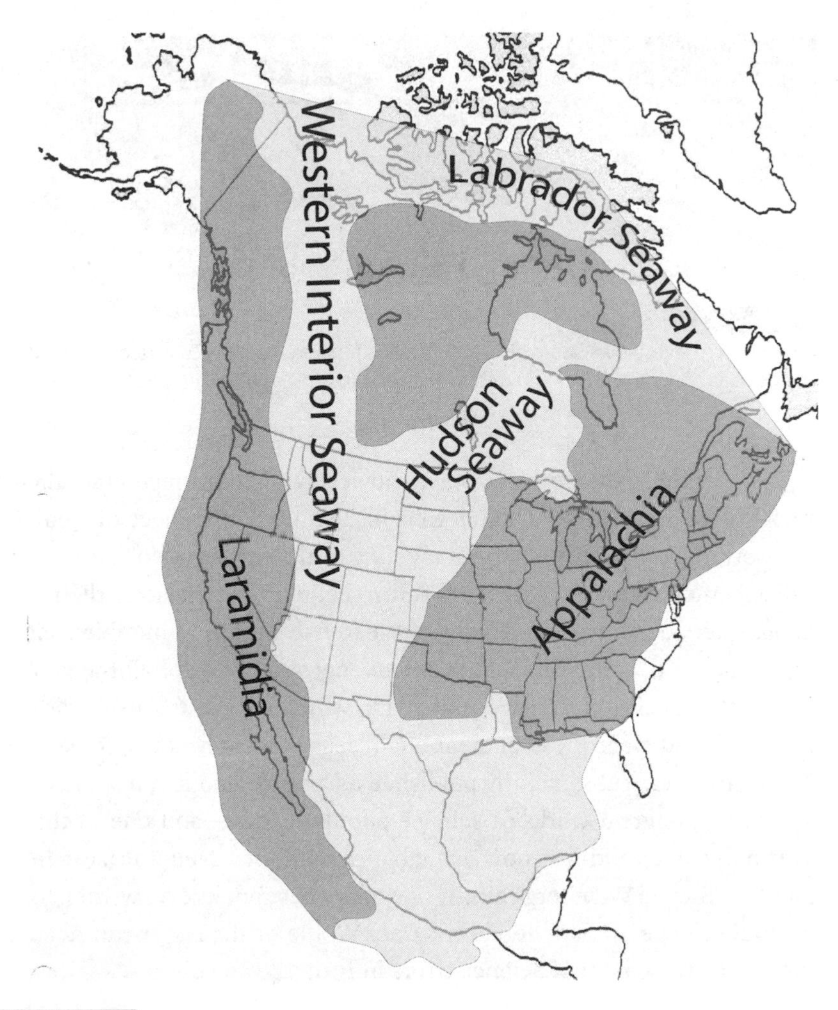

Figure 19.4 ▲

The Western Interior Cretaceous Seaway flooded the central plains of North America, from the Arctic Ocean to the Gulf of Mexico. (Redrawn from several sources)

Figure 19.5 ▲
Monument Rocks, Gove County, Kansas. Chalk is found not only in Europe, but in the Cretaceous beds of central Texas and western Kansas. (Courtesy of Wikimedia Commons)

None of these things were known, however, when the legendary English biologist Thomas Henry Huxley gave his lecture "On a Piece of Chalk" to the working men of Norwich at a meeting of the British Association for the Advancement of Science in 1868. A firm believer in educating the masses about science and nature, Huxley chose to use a piece of humble chalk on his chalkboard as the subject to spin an engrossing tale of all the geologic events that occurred to make that chalk, what kinds of fossils it was composed of, and what these ancient chalk seas must have once looked like. The lecture was subsequently published as a book, and it still stands today as one of the great works of science popularization—and one of the first ever published (and it is now available free online). Nobel Prize–winning physicist Steven Weinberg calls it "one of the best books ever written for the general science reader." Reviewer Dael Wolfle of the American Academy for the Advancement of Science wrote in 1967:

> That the lessons of paleontology are now so much more widely appreciated than they were when Huxley drew them from a piece of carpenter's chalk is

in good measure a tribute to Huxley's genius. We have much more factual knowledge than he had, but we have no better exemplar of the art of explaining in compelling and understandable terms what science is about, nor a more vigorous example of the scientist's obligation to practice that art.

FOR FURTHER READING

Everhart, Michael J. *Oceans of Kansas: A Natural History of the Western Interior Sea.* Bloomington: Indiana University Press, 2005.

Huxley, Thomas Henry. *On a Piece of Chalk.* New York: Scribner's, 1967.

Skelton, Peter W., Robert A. Spicer, Simon P. Kelley, and Iain Gilmour. *The Cretaceous World.* Edited by Peter W. Skelton. Cambridge: Cambridge University Press, 2003.

Smith, Andrew B., and David J. Batten, eds. *The Palaeontological Association Field Guide to Fossils, Fossils of the Chalk.* 2nd ed. London: Wiley-Blackwell, 2002.

20 THE DEATH OF THE DINOSAURS

THE IRIDIUM LAYER

The Age of Reptiles ended because it had gone on long enough and it was all a mistake in the first place. A better day was dawning at the close of the Mesozoic Era. There were some little warm-blooded animals around which had been stealing and eating the eggs of the Dinosaurs, and they were gradually learning to steal other things, too. Civilization was just around the corner.

—WILL CUPPY, *HOW TO BECOME EXTINCT*

SERENDIPITY

Thanks to generations of oversimplified science textbooks, most people think that science is all about planning your research to achieve some specific goal, making simple predictions, and then finding an answer through experiments. What most textbooks (or people) don't recognize is the element of luck in science. Many of the greatest discoveries in science happen not by planning, but by stumbling upon some unexpected result. This discovery by fortunate accident is often called "serendipity," after the old Persian tale of "The Three Princes of Serendip," who make discoveries unexpectedly.

There are hundreds of cases of accidental discoveries in science, especially in chemistry. Alfred Nobel accidentally mixed nitroglycerin and collodium ("gun cotton") and discovered gelignite, the key ingredient for his development of TNT. Hans von Pechmann accidentally discovered polyethylene in 1898. Silly Putty, Teflon, Superglue, Scotchgard, and rayon were all accidents, as were the discoveries of the elements helium and iodine. Among drugs, penicillin, laughing gas, minoxidil for hair loss, the Pill, and LSD were all discovered by accident. Viagra was originally developed to treat blood pressure, not impotence. Many of the great discoveries in physics and astronomy were unexpected, including the planet Uranus, infrared radiation, superconductivity, electromagnetism, X-rays, and many others. Among practical inventions, inkjet printers, cornflakes, safety glass, Corningware, and the vulcanization of rubber were all accidents. Percy Spencer was looking for another use for the surplus magnetrons after World War II ended and accidentally discovered that they could be used as microwave ovens when one of them melted the candy bar in his lab coat pocket. In 1964 two engineers, Arno Penzias and Robert W. Wilson, were trying to get the "noise" out of their newly developed microwave antenna. After getting rid of the routine "bugs," they found a background "hiss" that could not be eliminated. Even more surprising, the source of the noise was 100 times stronger than expected, and it was evenly spread across the sky, so it was not from a single point source on Earth or in space. Eventually, they realized they had discovered the long-predicted cosmic background radiation from the Big Bang. In 1978, they won the Nobel Prize for their discovery.

These and many other examples are a good reason why it is essential that science conduct "pure research" just for the sake of exploring and knowing things. Sadly, many shortsighted and misguided people (especially members of Congress trying to cut federal funding of science) scorn "pure research" as worthless navel-gazing and demand that every scientist show a practical or useful reason for his or her research or it will not be funded. This is a sure path to scientific stagnation. Even many of the scientific funding agencies operate this way, tending to reward research that is conventional and "more of the same" and seldom funding research that is a speculative gamble. Again and again, talking heads on television or politicians ridicule "pure research" that doesn't have a specific practical goal or application. Occasionally, narrow-minded and poorly educated people manage to interfere with the well-established scientific review process and shut down research they don't like (even though it was approved by legitimate scientists).

The sad irony of this entire misconception that "science must be practical and useful" is that most of the greatest discoveries in science are not anticipated or planned, but happen by accident. More often than not, scientists who found a crucial new piece of evidence were not looking for it, but for something else, and made their great discovery quite by accident. However, in the case of science, serendipity works most often when the researcher is prepared to see the implications of some new, unexpected development. Louis Pasteur put it this way, "In the field of observation, chance favors only the prepared mind." As noted scientist and writer Isaac Asimov said, "The most exciting phrase to hear in science, the one that heralds new discoveries, is not 'Eureka!,' but 'That's funny . . .'"

AN ACCIDENT IN THE APENNINES

A classic case of serendipity and accidental discovery is the discovery of the evidence of the event that ended the Age of Dinosaurs. For decades, there had been a pointless and inconclusive debate about what killed the dinosaurs at the end of the Cretaceous. Some said the climate got too hot; others said it was too cold. Some blamed it on the evolution of flowering plants—except that occurred in the Early Cretaceous, 80 million years earlier, and actually might have helped spur the evolution of herbivorous dinosaurs like the duckbills and horned dinosaurs. Some suggested that mammals ate their eggs—except that both mammals and dinosaurs originated together in the Late Triassic, about 200 Ma, and coexisted for 135 million years without mammals suddenly developing a taste for eating all the dinosaur eggs. Then there were even wilder and less scientifically testable ideas—epidemics and diseases, widespread depression and psychological problems, and even the notion touted in the tabloids that aliens kidnapped them or killed them off!

As paleontologist Glenn Jepsen wrote in 1964:

> Why become extinct? Authors with varying competence have suggested that dinosaurs disappeared because the climate deteriorated (became suddenly or slowly too hot or cold or dry or wet), or that the diet did (with too much food or not enough of such substances as fern oil; from poisons in water or plants or ingested minerals; by bankruptcy of calcium or other necessary elements). Other writers have put the blame on disease, parasites, wars, anatomical or metabolic disorders (slipped vertebral discs, malfunction or imbalance of hormone and endocrine systems, dwindling brain and consequent stupidity, heat

> sterilization, effects of being warm-blooded in the Mesozoic world), racial old age, evolutionary drift into senescent overspecialization, changes in the pressure or composition of the atmosphere, poison gases, volcanic dust, excessive oxygen from plants, meteorites, comets, gene pool drainage by little mammalian egg-eaters, overkill capacity by predators, fluctuation of gravitational constants, development of psychotic suicidal factors, entropy, cosmic radiation, shift of Earth's rotational poles, floods, continental drift, extraction of the moon from the Pacific Basin, draining of swamp and lake environments, sunspots, God's will, mountain building, raids by little green hunters in flying saucers, lack of standing room in Noah's Ark, and *palaeoweltschmerz*.

Without any independent evidence to test these ideas, they were just speculation, not science. Moreover, they all focused too much on dinosaurs and ignored the much more important picture: the end-Cretaceous extinction was a global event that affected the marine food chain (especially certain kinds of plankton and many kinds of marine animals) and plants on land. Any extinction event this widespread and pervasive needed a broader explanation than one specific to the dinosaurs. In fact, if it was this far-reaching and killed off so many other organisms at every level in the food chain, the dinosaur die-off is just an aftereffect, not the most important piece of the puzzle.

This was the state of the research into the Cretaceous extinctions when a young geologist named Walter Alvarez was doing field geology in the Apennine Mountains of central Italy. (I first met Walter in 1976 at Lamont-Doherty Geological Observatory when he was an unknown independent researcher and I was a graduate student.) His research focus had nothing to do with dinosaurs. He had long been interested in the structural geology of these rocks, and how they had been tilted and folded. As he mapped and described the thick sections of limestones that spanned the latest Cretaceous and early Cenozoic (Paleocene) near Gubbio, Italy, he noticed something unusual. Right at the boundary between the Cretaceous and the Cenozoic, there was a distinctive layer of dark clay instead of limestone (figure 20.1). This was then known as the "KT boundary," since the standard geological abbreviation for Cretaceous is "K" (from the German *Kreide* for "chalk"—see chapter 19); "T" stood for Tertiary, the interval of the Cenozoic from 66 Ma to 2.4 Ma. Since that time, geologists have tried to drop "Tertiary" as an obsolete term and use "Paleogene" for the interval

Figure 20.1 ▲
A close-up of the KPg boundary in Gubbio, Italy. The coin is on the boundary clay layer with the high iridium content. Below it are the white Cretaceous limestones, and above it are the postextinction beds of the Paleogene. (Courtesy of Wikimedia Commons)

from 66 Ma to 23 Ma instead. Thus, it is no longer the "KT boundary," but the "KPg boundary."

Then serendipity struck. Out of curiosity, Walter decided to see if the clay layer would give any clues about the duration of the KPg mass extinction. He took samples home to his new job at University of California–Berkeley, and consulted with his father, Berkeley physicist Luis Alvarez, about ways to use the clay layer to tell how long the extinction event took place (figure 20.1). Luis was already famous in the physics world, having worked on the Manhattan Project that built the atomic bomb and also having won the Nobel Prize for some of his own discoveries. The Alvarez father and son team decided that the clay might tell them something if they could detect particles of cosmic dust. A small amount of cosmic dust might suggest that the clay was deposited rapidly; a large amount would indicate that it had accumulated over a long time.

How do you measure ancient cosmic dust? Luis looked for trace elements that were extremely rare in crustal rocks but slightly more common in cosmic dust and other extraterrestrial materials. He focused the rare element iridium, a heavy metal near the bottom of the periodic table in the platinum group of metals. So they sent their samples to Berkeley physicists

Frank Asaro and Helen Michels, who ran the neutron activation analysis facility at Berkeley, where tiny amounts of trace material can be measured.

When the results came back, they were all shocked. The iridium amounts were off the charts! It was much more iridium than could be expected from just a long-term accumulation of cosmic dust. Then they tried to brainstorm an explanation for this anomalously high amount of iridium. If iridium came mostly from outer space, then it suggested something extraterrestrial was the source. They tried all sorts of ideas, from comets to many other possible explanations.

Finally, they calculated that it came from the impact of an asteroid about 10–15 kilometers (6–9 miles) in diameter that had hit the earth at the end of the Cretaceous. It would have had the energy of 100 million megatons of TNT, over a billion times the energy of the atomic bombs dropped on Hiroshima or Nagasaki (which Luis Alvarez actually witnessed as a scientific observer on a second B-29 bomber accompanying the *Enola Gay* to Hiroshima). Such an impact would not only scatter cosmic debris around the earth, but more importantly, it would fill the atmosphere with a cloud of dust that would produce the effects of a "nuclear winter," blocking sunlight and killing the plants on land and in the ocean and disrupting the food chain at its base. All of these ideas were put together and finally published in 1980 in a paper in the leading journal *Science*. The Alvarez, Alvarez, Asaro, and Michel paper has since become one of the most cited papers in the history of science.

THE IMPACT OF THE IMPACT

Naturally, when such an outrageous idea is first proposed to geologists, their reaction is to be skeptical. In all the sciences there are many provocative and controversial ideas that get through the first round of screening and peer review to be published, only to be shot down after the problem is re-investigated and more data collected. Scientists know from bitter experience not to take seriously every new discovery that is being hyped in the media, because most of them turn out not to be true, or at least not as remarkable as the press makes them out to be. Sadly, the press is a victim of its sensationalist "if it bleeds, it ledes" culture. They only report the eye-catching news once, then never get around to interviewing scientists who question the story, let alone reporting it has been debunked a few years later.

But at professional scientific meetings, hard evidence and not sensational claims rule the day. For many years after the paper by Alvarez and colleagues appeared, the program of the large scientific meetings (such as the Geological Society of America, or GSA, which I have attended every year without fail since 1978) were dominated by whole sessions arguing about the impact hypothesis, proposing new data, and debunking it. The claims seesawed back and forth, giving the neutral geologist whiplash trying to evaluate it all. At first geologists were skeptical about the iridium in the Gubbio clay layer, because clays are notorious for soaking up all sorts of rare components. But then anomalously high amounts of iridium were found in Stevens Klint, Denmark, and in deep-sea core sediments, so the iridium was not a local effect. But was it just in the oceans? If so, did it have an oceanographic geochemical cause? Then it was found in land sections in the Hell Creek beds of Montana, so we knew it was global and came out of the atmosphere to cover the earth. Even so, the difficulties of analyzing for iridium were manifest when one of the samples of high iridium came from the platinum in a lab tech's wedding ring—a single platinum-gold ring has much more iridium than do any of the samples from the KPg boundary.

By the early 1980s, there was pushback from another source. Many geologists had long known that the second-largest volcanic eruptions in the earth's history, the Deccan lavas of India and Pakistan, happened right around the KPg boundary. These could have released huge amounts of dust and ash into the atmosphere, producing an effect similar to the hypothetical "asteroid nuclear winter." It got even more interesting when it turned out that mantle-derived volcanoes like Kilauea also have lots of iridium in them. (Iridium is extremely rare in the crust but slightly more abundant in the mantle and in space.) Again the pendulum swung. The discovery of impact spherules (blobs of crustal matter from the crater), shocked quartz (only known from impacts and nuclear explosions), and giant tsunami deposits around the Caribbean and Gulf of Mexico pointed to the impact being real. Yet the improved dating of the Deccan lavas showed that they too were a big event that happened just before the KPg boundary.

The biggest stumbling block was the absence of the "smoking gun"—the KPg impact crater. Several candidates, like the Manson Crater in Iowa, were proposed and then rejected because new dates on them put them at the wrong age. The problem was finally solved—again by accident. Back in the late 1970s, an oil geologist named Glen Penfield had found geophysical data that showed a huge filled-in crater-like feature buried beneath the jungles of the northern Yucatan Peninsula in Mexico. But when he found it and published it in 1978 in an oil company report, no one was interested in the asteroid impact model yet. A decade later planetary geologist Alan Hildebrand realized that all the tsunami deposits and impact droplets were scattered around the Caribbean and Gulf of Mexico, and he began to look for the crater there. He found Penfield's reports in 1990. Since then, the buried crater known by its Mayan name as Chicxulub (CHIK-zoo-loob) has been drilled and studied and dated precisely at the KPg boundary, confirming it as the impact site.

WHAT DO THE FOSSILS SAY?

It's all well and good to establish that a big impact happened at the end of the Cretaceous, and many geologists (especially non-paleontologists) stopped right there and declared it "case closed." But it's more complicated than that, since we know the gigantic Deccan eruptions were happening at the same time, even starting slightly before the impact happened, so how do you tease the two causes apart? Adding to this complication is the fact that there was a giant sea-level drop at the end of the Cretaceous, draining and exposing huge areas of inlands seas like the chalk seas that once covered Europe and the Western Interior Seaway in the Great Plains of North America (see chapter 19). This would have a giant effect on marine animals that depended on all that shallow seafloor for their habitats.

The best arbiter of this complicated nexus of causes would be the fossil record. After all, it was the mass extinction of dinosaurs, ammonites, and many other creatures that led us to look for the cause of the KPg extinction in the first place. So that's where a lot of research has been focused. If the victims of the KPg extinction all died out simultaneously at the horizon with the iridium and the other impact deposits, then the impact scenario would prevail. If, however, they were slowly dying out through the latest Cretaceous, or died before the impact, or survived the impact and then died off, then the slow effects of long-term climate change from the Deccan eruptions, and possibly the sea-level change, would be more important.

And that is where most of the argument and data collection has been focused for the past 37 years since the original Alvarez impact extinction model was first published. To cut to the punch line, the pattern is not a

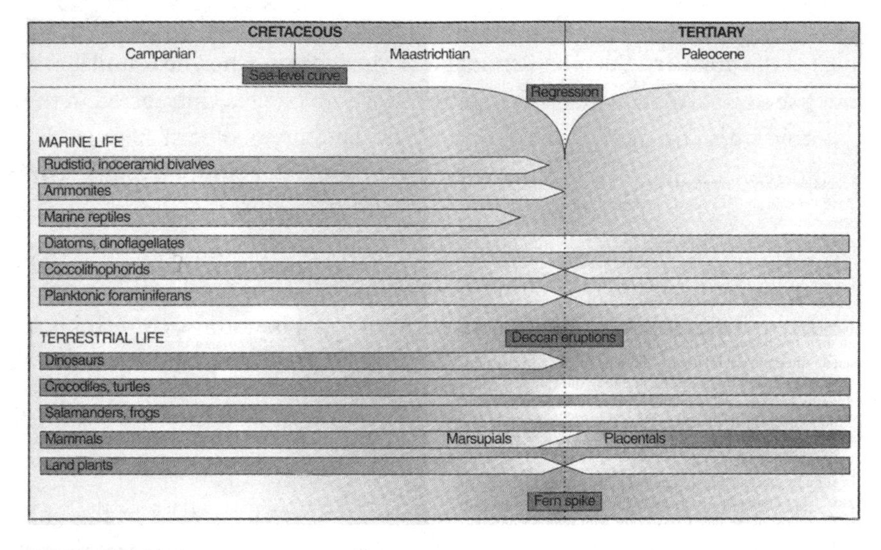

Figure 20.2 ▲

The pattern of extinction and survival across the KPg boundary. (Original drawing by the author)

simple one where everything goes out at the iridium level (figure 20.2). Only a few groups of animals actually died out at the impact horizon, and a surprising number survived the impact, or died out or declined long before it happened.

In the marine realm, there are two groups of plankton that seem to have crashed at the time of the impact (the amoeba-like foraminifera and the coccolithophorid algae), and they were really the only significant victims among the plankton (see figure 19.2). By contrast, three other groups of plankton (the algae known as diatoms and silicoflagellates, plus the amoeba-like radiolarians) registered little or no effect. Most of the marine invertebrates (sponges, corals, sea urchins, sea lilies, brittle stars, brachiopods, and bryozoans) were minimally effected or not effected at all. Two groups of huge clams (the dinner plate–shaped inoceramids and the cone-shaped colonial rudistid reefs) disappeared long before the impact, apparently due to the effects of the Deccan volcanic gases on the oceans. Of the remaining molluscs, 35 percent of snails and 55 percent of clams and oysters died out, but every study shows their extinction was gradual through the end of the Cretaceous. The only marine invertebrates that died out near or at the KPg boundary were the ammonites, and most paleontologists agree that they were declining long before the end of the Cretaceous and may not have even been alive when the rock from space arrived. In some places, like Antarctica, they gradually declined through the latest Cretaceous until almost none were left at the KPg boundary, consistent with the deteriorating climate due to the Deccan volcanic gases. In addition, the marine reptiles (mosasaurs, plesiosaurs, and giant turtles) were also in decline in the latest Cretaceous, and there is no clear evidence they were alive to witness the impact.

On land, the pattern is equally complex. There is a signal from the pollen that the typical latest Cretaceous flora (the *Aquilapollenites* pollen assemblage) died out at the iridium horizon, and there is a striking abundance of fern spores right at the KPg boundary, consistent with a time of plant death and cooling and darkness. But the rest of the land animals give complex and contradictory responses. Sure, dinosaurs (other than their bird descendants) died out, but the latest research shows they were in decline long before the KPg boundary, and only a few individuals of *Triceratops* and *Tyrannosaurus* may have survived to see the fireball slam into earth. But nearly all the other land animals (crocodilians, turtles, snakes, lizards, freshwater fish, and frogs and salamanders) made it through the supposed hellish "nuclear winter" on Earth with almost no extinctions. If the world were so extreme as some versions of the impact model claim, how did crocodilians survive and some dinosaurs that were smaller than crocodilians die out? Some of them may have survived in aquatic habitats during the worst of the firestorm, as has been argued, but not for very long. Others have suggested that crocodilians hibernated in burrows in riverbanks as some do today. But they need long preparation time in order to hibernate for winter, which the impact would not have given them—if indeed they hibernated at all. Remember, Cretaceous winters were quite mild, since it was a greenhouse world with almost no snow or ice anywhere at the time. The mammals showed a shift in dominance from the opossum-like marsupials to some of the first dominant placental mammals groups in the Paleocene, but relatively little overall drop in diversity. The squirrel-like egg-laying multituberculates vanished in China, but not in North America. Finally, any extreme scenario that suggests huge amounts of sulfuric acid rain from the impact hitting the gypsum beds of Yucatan ignores the fact that amphibians cannot tolerate even small amounts of acidity in their habitat, because of their porous skins. Even the slight changes in acidity happening today due to modern acid rain are causing havoc among frogs and salamanders.

META-ANALYSES

In short, the fossil record does not support a simple "the rock from space did it, end of story" scenario that so many reporters and people have heard (and some scientists also believe). Within the scientific community, the debate still goes on, more than 37 years after it began, and shows no signs of slowing down. Every year I attend the national GSA meeting in different cities in the fall, and every year there are more sessions arguing the details of new data. For a while, it appeared that the impact advocates had prevailed, but in the 2014 GSA meeting in Vancouver, the tide was turning back to the Deccan volcanoes, and the same was true at the 2015 GSA meeting in Baltimore and the 2016 meeting in Denver.

What is clear is that three things happened about the same time: impact, eruptions, and sea-level drop. All three must have contributed, and no single cause is sufficient to explain all the effects. Nature is complex and defies simplistic models; there is no simple "right" answer concerning such a complex event as why the KPg extinctions occurred, no matter how much the media want to oversimplify it so it is easy to write about and keep within their word limits.

Meanwhile, the never-ending debate is polarized along the lines of specialization. Geologists, geophysicists, and geochemists tend to like simple, clear-cut answers once they get their data out of their machines, and they tend to favor the impact-only scenario. Paleontologists, on the other hand, are trained in biology, and recognize the complexity of living systems that defies simplistic answers. Vertebrate paleontologists, who study the dinosaurs, reptiles, amphibians, and mammals, were polled in 1985, and only 5 percent agreed that the impact was the cause of the KPg extinctions. In 1997, a survey of 22 distinguished British paleontologists (specialists in each of the groups that lived in the Late Cretaceous) voted overwhelmingly against the impact being significant in the marine fossil record. In 2004, a survey of vertebrate paleontologists found only 20 percent accepted the impact cause for the KPg extinctions, whereas 72 percent felt that it was a gradual process consistent with the Deccan eruptions and not the impact. In 2010, a paper written by multiple authors (few of whom were paleontologists) was published in the eminent journal *Science*, again pushing the impact as the sole explanation of the KPg extinctions. It was immediately rebutted by a paper authored by 28 paleontologists that demonstrated that the impact was only a minor part of the story. Even Walter Alvarez, in his popular book *T. Rex and the Crater of Doom*, conceded that the KPg extinctions had multiple complex causes.

In short, the battle shows no sign of abating and is largely polarized along the lines of professional specialization. But even more is at stake here. Some people have built their entire careers around promoting one model or the other, so they have a lot to lose: grant money, publications, prestige, and even personal pride. They are unlikely to back down, no matter what the evidence. Scientists are human, after all, and if the verdict doesn't force them to concede, they won't do so voluntarily.

Sometimes it gets even more personal than this, and there was certainly a lot of name-calling and career-wrecking happening during the rough-and-tumble years of the debate in the 1980s and 1990s. Luis Alvarez said: "I don't want to say bad things about paleontologists, but they're really not very good scientists. They're more like stamp collectors." On the opposite side, dinosaur paleontologist Bob Bakker told a reporter:

> The arrogance of these people is simply unbelievable. They know next to nothing about how real animals evolve, live, and become extinct. But, despite their ignorance, the geochemists feel that all you have to do is crank up some fancy machine and you've revolutionized science. The real reasons for the dinosaur extinctions have to do with temperature and sea level changes, the spread of diseases by migration and other complex events. In effect, they're saying this: we high-tech people have all the answers, and you paleontologists are just primitive rockhounds.

With attitudes like this, and so much at stake, not only will the answer be the muddled, "It's complicated," but the debate will never end until all the original combatants have left the field through death or retirement. And that day has not yet come.

Whatever its scientific fate, the discovery of the Gubbio iridium anomaly and the proposal of the Alvarez asteroid extinction model have been good for science. The arguments produced huge amounts of new detailed research and thousands of scientific papers and dozens of books.

The debate breathed new life into certain branches of geology and paleontology and launched many careers (but it destroyed a few as well). For a long time, it led scientists to look at geology in a different way, emphasizing the rare catastrophic but natural event that the extreme gradualism of Charles Lyell had long suppressed. It probably went too far at one point. During the 1980s and 1990s, some scientists tried to blame all extinction on impact events, only to find out that none of the other the extinction horizons show any evidence of impact—only the KPg. But that's how science operates. We make mistakes, but sooner or later peer review and lots of additional research fixes them, and we get the right answer. And the world is richer for it.

FOR FURTHER READING

Alvarez, Walter. *T. Rex and the Crater of Doom*. Princeton, N.J.: Princeton University Press, 1997.

Archibald, J. David. *Dinosaur Extinction and the End of an Era: What the Fossils Say*. New York: Columbia University Press, 1996.

——. *Extinction and Radiation: How the Fall of the Dinosaurs Led to the Rise of the Mammals*. Baltimore: Johns Hopkins University Press, 2011.

Dingus, Lowell, and Timothy Rowe. *The Mistaken Extinction: Dinosaur Evolution and the Origin of Birds*. New York: Freeman, 1997.

Keller, Gerta, and Andrew Kerr, eds. *Volcanism, Impacts, and Mass Extinctions: Causes and Effects*. Geological Society of America Special Paper 505. Boulder, CO: GSA, 2014.

Keller, Gerta, and Norman McLeod. *Cretaceous-Tertiary Mass Extinctions: Biotic and Environmental Change*. New York: Norton, 1996.

Officer, Charles, and Jake Page. *The Great Dinosaur Extinction Controversy*. New York: Helix, 1996.

Powell, James Lawrence. *Night Comes to the Cretaceous: Dinosaur Extinction and the Transformation of Modern Geology*. New York: St. Martin's, 1998.

21 | HOW PALEOMAGIC LAUNCHED PLATE TECTONICS

LODESTONES

Magnus magnes ipse est globus terrestris. (The Earth itself is a great magnet.)

—WILLIAM GILBERT, *DE MAGNETE*

PUZZLE #1: LODESTONES AND EARTH'S MAGNETISM

Since ancient times, people had been mystified by rocks they called lodestones (now known to be chunks of the mineral magnetite) (figure 21.1). As early as the sixth century B.C.E., the Greek philosopher Thales of Miletus described how these peculiar rocks were attracted to each other and how small pieces of iron stuck to them. There are references to magnetic rocks in the fourth century B.C.E. in the Chinese book *Guiguzi* (Book of the devil valley master). In the second century B.C.E., the Chinese chronicle *Lusi Chuqiu* read: "A lodestone makes iron come or attracts it." The *Lunheng* (Balanced inquiries), written sometime between 20 and 100 C.E., noted: "A lodestone attracts a needle." By the twelfth century, Chinese navigators had figured out how to float a chip of lodestone on a cork to make a crude compass. And in 1190 C.E., William of Neckham described a lodestone compass, indicating such compasses were in widespread use not only in China but also in Europe by that time.

Speculations about what mysterious force caused the lodestone to seek the north and attract other lodestones or pieces of iron continued for centuries. In fact, the word "magnetism" came to be applied to any unexplained force that acted at a distance, originally without any real

Figure 21.1 ▲
This version of a lodestone compass invented by Galileo suspended a natural piece of magnetite on a wire that pivoted and turned freely so the device could be used to determine magnetic north. (Courtesy of Wikimedia Commons)

connection to what we now know about magnets or electricity. Even today, we still talk about "animal magnetism" or people with "magnetic" personalities. But in 1600, the English natural philosopher and physician William Gilbert published a book written in Latin (the language of scholars back then), entitled *De Magnete*, that summarized almost everything known about magnetism and magnets at the time. His work is considered the beginning of the modern understanding of electricity and magnetism. He correctly concluded that there were invisible fields around a bar magnet that attracted other magnetic objects and that could be seen by sprinkling iron filings around the magnet. He also argued that the earth must be a huge magnet, which explained why lodestones always pointed north. In fact, he was far ahead of his time in arguing that there must be a big mass of iron in the earth's interior, something that was only confirmed much later by seismology and studies of gravity and meteorites (see chapter 10).

Gilbert also correctly argued that the earth rotates on its axis and hinted at support for the Copernican heliocentric solar system (first published in 1543), even though that idea was still considered heretical in most of Christendom. Remember, this was still 20 years before Galileo championed the heliocentric system and the Inquisition threatened him with torture if he didn't recant his heresy. Gilbert pointed out that the idea of "fixed stars" on a great dome of the heavens (as was commonly believed then) was absurd, along with the "celestial spheres" on which the planets moved outside that dome. Instead, he realized that the stars were points of light from sources located immense distances away from us. He also dabbled in the nature of static electricity. He even coined the word "electron" from the Latin word *electrum*, meaning "like amber," because it is easy to generate static by rubbing a piece of amber on fabric.

Gilbert never had the chance to follow up on his brilliant work because he died of the bubonic plague only 3 years after it was published. But for centuries afterward, the idea of magnetic lodestones, along with further breakthroughs in the understanding of electricity, was one of the cutting-edge areas of research in science. In the early 1800s, Michael Faraday had done many experiments that demonstrated the nature of the earth's magnetic field and the connection between magnetic fields and electrical fields. In the 1860s, James Clerk Maxwell unified electromagnetism with his brilliant mathematical explanations of Faraday's experiments.

The source of the earth's magnetic field was long a mystery, subject to many myths. Early philosophers and naturalists assumed the source of the field was in the heavens, because it was measured through the air with no obvious physical cause. During the time of the Greeks and Romans, the field was thought to be caused by some huge magnetic mountain, somewhere to the far north. The Hellenistic Greek philosopher and astronomer Claudius Ptolemaeus of Alexandria (better known as the Ptolemy who is famous for his system of epicycles to explain the problems with the orbits of the earth-centered universe) described a legendary island near Borneo that was so magnetic that it held ships in place by attracting the nails in their boards. In the famous Arabian tales of Aladdin and Sinbad the Sailor in *One Thousand and One Nights*, there is a legend of a mountain whose magnetism was so strong it pulls a ship's nails out of its boards and causes it to fall apart and sink. The famous mapmaker Mercator placed a mountain of iron on the North Pole, and when that didn't work to explain the differences in magnetic direction due to the difference between the North rotational pole and the North Magnetic Pole, he put two mountains of iron at the top of the map.

The true cause of the earth's magnetic field was not discovered until the middle of the twentieth century, when geologists concluded that the earth's core was made of a dense iron-nickel alloy (see chapter 10). The temperature of the earth's core is over 4,000°C, too hot to have a permanent bar magnet in the middle. (Solid bar magnets are demagnetized if they are heated above 650°C.) In 1946 Walter Elsasser first suggested how movements of the iron fluids in the core might generate the observed magnetic field of the earth. This was followed by the work of physicist Edward Bullard, who first modeled the earth's magnetic field and core movements mathematically by using geophysical fluid dynamics.

Since that time, geophysical fluid modeling by Elsasser, Bullard, and many others has shown how magnetic dynamos spin inside the fluid iron-nickel outer core of the earth. Just as the dynamo in the hydroelectric plant in a dam spins a coil of conducting wire around a magnet to make electricity, so the geodynamo spins a mass of conducting iron-nickel through the magnetic field of the earth, generating an electrical current—which, in a great feedback loop, generates a stronger magnetic field, a stronger current, and so on. The actual mathematical details of how this works is very complicated, but it is the only explanation that fits the known properties of the earth's magnetic field.

PUZZLE #2: POLAR WANDER CURVES

By the mid-1800s, scientists had invented crude devices for detecting the presence of a magnetic field in a solid sample of material, such as a rock. Eventually, these became more and more sophisticated, so that by the 1930s, there were proton-precession magnetometers that could detect the magnetic signature of a submarine beneath the waves. By the 1940s, there were fluxgate magnetometers that could detect the weaker magnetic fields in a rock sample.

In 1948 Ellis Johnson, Thomas Murphy, and Oscar Torreson of the Department of Terrestrial Magnetism at the Carnegie Institute of Washington published a very influential paper entitled "Pre-history of the Earth's

Magnetic Field." By analyzing the magnetic directions in layers of ancient sediment from a glacial lake, they showed that not only lavas and volcanic rocks could have a strong magnetic signal, but so could sediments. They were able to show the variations in the earth's field layer by layer, spanning the interval from 17,000 to 11,000 years ago. This paper spurred a number of scientists to look into the study of the earth's ancient magnetic field (paleomagnetism). Scientists who study paleomagnetism have been nicknamed "paleomagicians," because ancient magnetism has solved a wide range of interesting and challenging geologic problems.

This research caught the eye of the brilliant physicist Patrick Blackett, who had just won the Nobel Prize in 1948 for his investigation of cosmic rays using a cloud chamber. Born in London in 1897, Blackett served in the Royal Navy in World War I and survived several battles, while developing improvements in gunnery and other naval equipment. After the war, he moved on to physics at Cambridge University's famous Cavendish Laboratory. There he became a protégé of Ernest Rutherford, the pioneer of radioactivity. Blackett and Rutherford developed the cloud chamber and eventually discovered anti-matter. By 1947 Blackett was working on developing a magnetometer to detect the earth's magnetic field, hoping that it would help link electromagnetism with another fundamental force, gravity. This effort did not succeed, but in the process he developed better and better magnetometers and also collected a lot of paleomagnetic data on ancient rocks. Among Blackett's students at Cambridge was Keith Runcorn, who would pioneer a whole new field of paleomagnetism: looking at the change of the pole position recorded in the magnetic rocks of continents over millions of years. Runcorn, in turn, launched a big paleomagnetism program when he moved on to Newcastle University, training such eminent scientists as Ted Irving, D. W. Collinson, Ken Creer, Neil Opdyke, and many other famous researchers in the paleomagnetics field. (I was trained by Opdyke, so I'm a direct intellectual descendant of the Rutherford-Blackett-Runcorn lineage.)

Runcorn and his students and colleagues set out to compile as many paleomagnetic directions from as many different rocks of different ages in different continents as they could. By the mid-1950s, the data set was very large, and certain patterns were beginning to emerge. The first thing they could tell was that the position of any continent had changed over time with respect to the North Magnetic Pole. If you analyzed the magnetic directions of very young rocks, they all pointed to the modern magnetic north. But as you looked at older and older rocks on any given continent, the ancient magnetic north direction recorded in those older rocks moved farther and farther from modern magnetic north. From the perspective of any one continent, you could fit those magnetic directions to form a curve showing the motion of the North Magnetic Pole with respect to the continent. This was called "polar wander."

But complications arose when you did the same exercise with a second continent (figure 21.2). Again, the youngest rocks would have directions close to modern magnetic north, but older rocks would produce magnetic north directions that were farther and farther from the North Pole. Fitting a curve through all these data, you would get a second polar wander curve—and it did not match the one from the first continent. The same problem would occur with the data from a third continent. The youngest rocks matched modern magnetic north, but older rocks produced a polar wander curve that did not match those of other continents.

This created a dilemma. If you assumed the continents did not move (which most geologists still believed in the 1950s), then it was peculiar that each continent had a distinctly different polar wander curve that ended up in the same place today—the modern magnetic north. This implied that the earth had many different magnetic poles throughout its geologic past

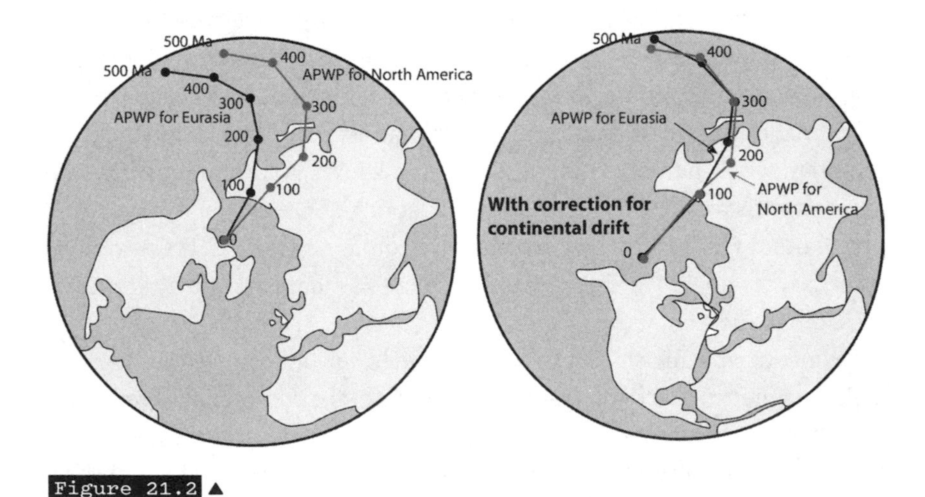

Figure 21.2 ▲

The plot of apparent polar wander curves (APWP) with continents in their fixed positions. If they are allowed to move, then the curves match exactly. (Redrawn from several sources)

and that they coincidentally happened to converge on the North Pole at the present time.

But what if you allow the continents to move instead? If you rotate them back to their original positions in the past, all of their different "polar wander" curves match beautifully. In other words, the North Magnetic Pole hasn't moved much from the earth's rotational pole over geologic time—but the ancient rocks recording the magnetic direction to that pole have moved with respect to it, giving an "apparent polar wander curve."

This was powerful evidence for continental drift, much of it published in the mid-1950s. And yet the mainstream geological community was not prepared to accept it or to toss out their long-held assumption of fixed continents. Most of them thought that the field of paleomagnetism was too new, and there were too many pitfalls in paleomagnetic data (both known and unknown at that time) to decide whether they really did show that the continents had moved. So geologists who knew of these studies simply took a "wait and see" attitude.

PUZZLE #3: THE EARTH'S MAGNETIC FIELD HAS FLIPPED!

In addition to apparent motions of the earth's magnetic field over time, there was another property of the magnetism in ancient rocks that proved to be surprising: magnetic directions flip every now and then. Today, for example, magnetic compass needles point north, but 800,000 years ago a compass's north arrow would point south. In other words, the field completely reverses itself every few thousand to few million years, so the field lines from the North Pole to South Pole change direction, and the earth has a different magnetic polarity. The magnetic directions in rocks that match today's magnetic north direction are called "normal" by convention, so those that point south are known as "reversed" in their polarity.

Rocks that had reversed polarity were first discovered by the French physicist Bernard Brunhes in 1905, through analyses of the same volcanoes in Auvergne, France, that had proven critical to the confirmation of plutonism in the 1770s (see chapter 5). This was just an isolated observation, however, so no one knew what to make of it. Not until 1929 did someone go further, when the Japanese geophysicist Motonori Matuyama collected over 100 basalt specimens from Japan and Manchuria and showed that they pointed in only two directions: modern magnetic north or 180° away (reversed polarity).

He also placed the rocks in a time sequence and could tell that nearly all the rocks of reversed polarity were roughly the same age and that there were other times when all the rocks were of normal polarity. By 1933 Swiss geophysicist and Arctic explorer Paul Mercanton had also documented reversed volcanic rocks as well as clays baked by the intrusion of volcanic rocks, and argued that they showed both normal and reverse polarity and might somehow be useful to testing Wegener's new continental drift hypothesis.

However, for the next 25 years the peculiarity of rocks with reversed magnetization was neglected while geophysicists focused on other things, including the polar wander idea. Part of the problem was the misleading behavior of a peculiar rock called the Haruna dacite from Japan. Analyzed in 1951, it had the odd property that it was self-reversing, giving one direction when you analyzed it initially, then changing polarity when you heated it, and then changing again each time it was analyzed. This strange rock made many scientists wary of claims that rocks had reversed magnetic directions, because it was not clear that those magnetic directions were primary directions that were locked in when the rock formed; they could all be behaving like the Haruna dacite. This glitch discouraged paleomagnetists for years from pursuing the problem.

Then, in the late 1950s and early 1960s, a trio of scientists decided to tackle the problem of magnetic reversal on a large scale: two Stanford paleomagicians, Allan Cox and Richard Doell, and a young Berkeley geochemist named G. Brent Dalrymple (figure 21.3), who had been a pioneer in refining the potassium-argon dating method. Shortly after getting his Ph.D. from Berkeley, Dalrymple took a job at the U.S. Geological Survey in Menlo Park, just down the road from Stanford. (Dalrymple was an alumnus of tiny Occidental College, where I taught for 27 years, so I know him.) Cox, Doell, and Dalrymple sought to test the two alternative explanations for reversely magnetized rocks: if reversed directions in ancient rocks was a problem with individual samples that self-reversed, then if you analyzed a bunch of samples of the same age around the world, they would show mixed polarity. However, if the crazy idea that the entire magnetic field of the earth had reversed were true, then rocks of the same age all over the world should give the same normal or reversed polarity.

Cox, Doell, and Dalrymple went on collecting trips all over the world and sampled as many different rock outcrops as they could find. These were mostly younger lava flows, which not only can be dated by

Figure 21.3 ▲

Allan Cox (*left, seated*), Richard Doell (*center, standing*), and G. Brent Dalrymple (*right*) in the U.S. Geological Survey magnetics lab in 1965. At the time, they were rapidly developing the magnetic polarity time scale.(Courtesy of U.S. Geological Survey)

potassium-argon methods, but also tend to be strongly and stably magnetized. Over the next decade, they analyzed hundreds of rock units and thousands of samples, with Dalrymple determining the potassium-argon age, and Cox and Doell doing the paleomagnetic analysis. By the early 1960s, they were beginning to see a pattern (figure 21.4A). The youngest rocks on the earth (younger than 780,000 years) were all normal in polarity. But a consistent pattern of normal and reversed polarity emerged as older and older samples were examined. Most of the rocks between 780,000 years ago and 2.5 Ma were reversed in polarity, with a few minor normal events interspersed with the long period of field reversal. Rocks from 2.5 to 3.4 Ma were mostly normally magnetized, then those older than 3.4 to about 5 Ma were a mixture of normal and reversed. There appeared to be a clear worldwide signal of changes in the polarity of the earth's magnetic field, and it occurred in every rock suitable for analysis no matter where it was located. Further, it was not a glitch caused by freakish rocks like the Haruna dacite.

Their research continued through the 1960s, uncovering more and more long and short polarity events across the past 10 million years. Their pace was spurred by a friendly competition with Ian McDougall (geochronologist) and Don Tarling and Francois Chamalaun (paleomagnetists) at the Australian National University in Canberra, who were tackling the same problem and trying to sample and analyze as many rocks as they could. By the late 1960s, the two labs established what is now known as the magnetic polarity time scale (figure 21.4B). If you could find the pattern of magnetic signal in a rock, you could match it to the history of magnetic reversals over the geologic past and get its age with high precision.

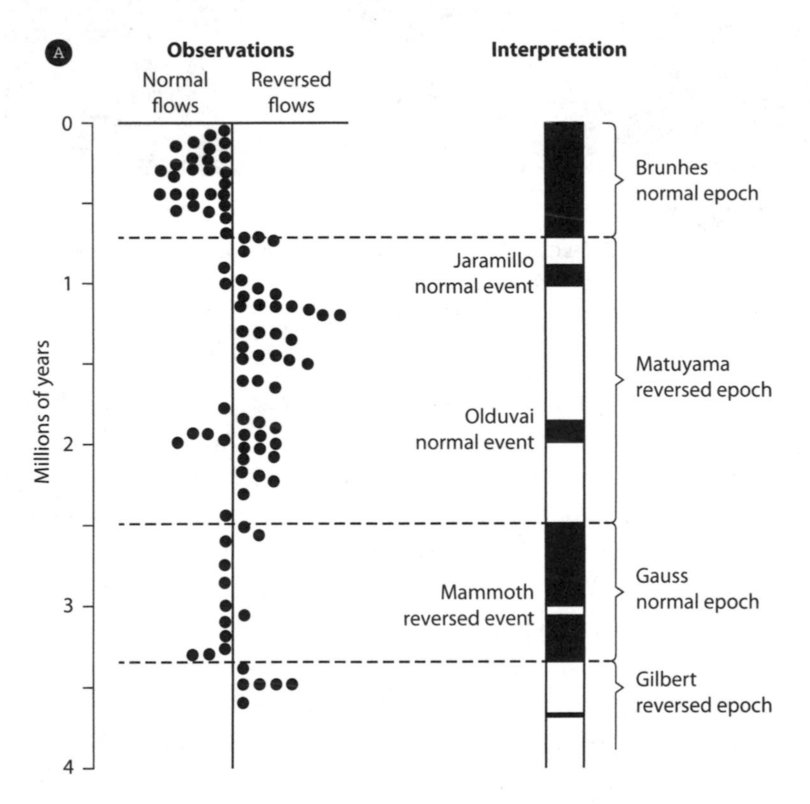

Figure 21.4 ▲

The magnetic polarity history of the earth: (*A*) Early magnetic time scale, showing dated lava flows that are all of the same polarity (normal or reversed) at the same age. (*B*) The past 4.5 million years of magnetic polarity history, as deciphered during the 1960s by Allan Cox, Richard Doell, and G. Brent Dalrymple of the U.S. Geological Survey, and Ian McDougall, Don Tarling, and Francois Chamalaun of the Australian National University. (Modified from several sources)

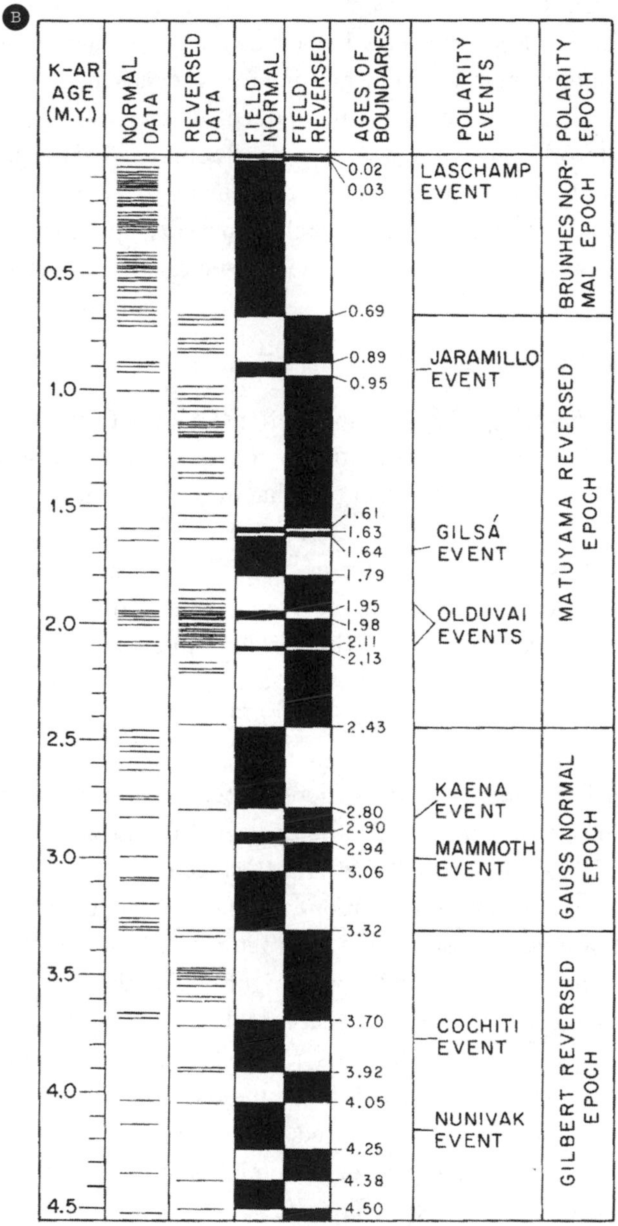

B

K-AR AGE (M.Y.) | NORMAL DATA | REVERSED DATA | FIELD NORMAL | FIELD REVERSED | AGES OF BOUNDARIES | POLARITY EVENTS | POLARITY EPOCH

0.5
1.0
1.5
2.0
2.5
3.0
3.5
4.0
4.5

0.02
0.03
0.69
0.89
0.95
1.61
1.63
1.64
1.79
1.95
1.98
2.11
2.13
2.43
2.80
2.90
2.94
3.06
3.32
3.70
3.92
4.05
4.25
4.38
4.50

LASCHAMP EVENT
JARAMILLO EVENT
GILSÁ EVENT
OLDUVAI EVENTS
KAENA EVENT
MAMMOTH EVENT
COCHITI EVENT
NUNIVAK EVENT

BRUNHES NORMAL EPOCH
MATUYAMA REVERSED EPOCH
GAUSS NORMAL EPOCH
GILBERT REVERSED EPOCH

Figure 21.4 ▲
(*continued*)

This was a giant breakthrough in geology that earned Cox, Doell, and Runcorn the Vetlesen Prize, the highest award in geology or geophysics. It led to an entirely new discipline of using the reversals in the magnetic field recorded in sediments on land and in deep-sea cores and provided a new method of worldwide dating with high precision. (Most of the grant-funded research I did during my career was done using this method of magnetic stratigraphy.) But little did Cox, Doell, and Dalrymple suspect that their data would also unlock the secret of seafloor spreading and plate tectonics.

PUZZLE #4: STRIPES ON THE SEAFLOOR

As soon as marine geology got going after the end of the World War II, it was routine for researchers to gather as much different data as they could, not knowing what they might find or what the data might mean. Every ship running out of Scripps, Lamont, or Woods Hole all over the world's oceans routinely collected data about the depth of the seafloor, the geology of the layers below the sea bottom, and the temperature, salinity, and chemistry of the seawater at various depths. Whenever possible, a long steel tube known as a piston corer would be dropped down into the ocean floor sediments to recover a core sampling the layers of sediment back through millions of years of ocean history.

Also on board these ships was a long torpedo-shaped instrument known as a proton-precession magnetometer. It had originally been developed to hunt submarines during World War II, since the steel hulls of ships are strongly magnetic. Once the war ended, many oceanographic institutions inherited war-surplus ships and magnetometers for marine geological research. These were routinely towed right behind the ship and took continuous readings of whatever magnetic fields they sensed.

In the late 1940s and early 1950s, a huge amount of magnetic data from the seafloor had been collected. Once the data were converted from old computer files, analyzed, and plotted, the results did not look like random noise, but showed a odd pattern: when plotted on a map, the data points created huge "stripes" of different magnetic strengths that ships had encountered while sailing over the seafloor (figure 21.5). These "stripes" were formally known as magnetic anomalies, because the measurements were anomalously different from the background value of the magnetic field that you feel all your life. In some cases, there were positive anomalies,

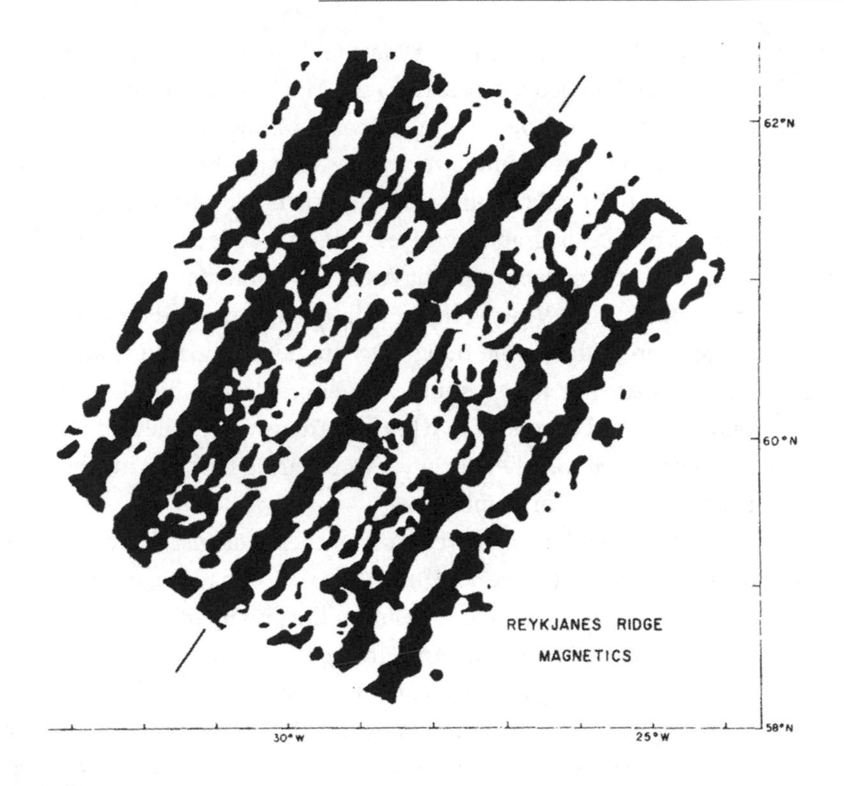

Figure 21.5 ▲

The magnetic polarity pattern of the seafloor. This is the original pattern over the Reykjanes Ridge, on the Mid-Atlantic Ridge south of Iceland, with bands of normal polarity (*black*) and reversed polarity (*white*) corresponding to the original positive and negative magnetic anomalies measured by the shipboard magnetometer. (Redrawn from several sources)

with readings stronger than the background value of the earth's ambient field. In other cases, the negative magnetic anomalies were weaker than the average magnetic field detected by the magnetometer.

Geologists were not sure what to make of this pattern. Was it alternating bands of different types of rock? Was it alternation of strongly and weakly magnetized rocks? By the late 1950s, Scripps geophysicists R. G. Mason and A. D. Raff had plotted the data for the Pacific seafloor off the coast of Alaska, British Columbia, Washington, and Oregon, but the pattern was complex and confusing. It was not apparent what caused these parallel magnetic stripes.

Meanwhile, a large group of geologists and geophysicists at Cambridge University were working on similar problems, supervised by physicist

Edward Bullard. In 1961 Drummond Matthews (figure 21.6A) completed his Ph.D. based on dredged samples from the Mid-Atlantic Ridge. His work showed it was entirely made of similar-looking flows of basaltic lava, so there was no real difference in the rock types that might explain the magnetic stripes. In 1962 another young student, Frederick Vine, began to work with Matthews on oceanographic cruises, not only collecting samples, but also looking at the big collection of magnetic data obtained from the same part of the seafloor. First, Vine worked on data Matthews had brought back from the Carlsberg Ridge in the Indian Ocean, but in 1962, Vine and Matthews started looking at the existing magnetic data for the Mid-Atlantic Ridge south of Iceland (figure 21.5), where Matthews had collected his rock samples. They worked together closely, even sharing lodgings in what had been an old set of stables in Cambridge.

They were also inspired by the world of another geologist, Harry H. Hess of Princeton University (figure 21.6B). In 1931 Hess had been a scientist on a U.S. Navy submarine voyage measuring the gravity of the ocean floor from Florida and Cuba to the eastern Caribbean. During World War II, he had been captain of a Navy vessel, an attack transport named the USS *Cape Johnson*, cruising back and forth across the Pacific. As captain, he kept the echo sounder always turned on day and night, so he was continually collecting data about the seafloor beneath his vessel.

In the process, he was the first to discover many key features of the seafloor, including huge seamounts that towered above the deep-sea bottom. Some of these seamounts had flat tops. As their discoverer, he was entitled to name them, and he called them "guyots," after Arnold Guyot (pronounced "gee-YOH"), the Swiss geologist who had founded the Princeton Geology Department (which still resides in Guyot Hall).

When the war ended, Hess remained in the Naval Reserve for some time and rose to the rank of rear admiral, even as he resumed his teaching duties at Princeton. He began to compile all the information coming in from the various oceanographic institutions over the previous decade. From his Navy work, he knew that seamounts seemed to be volcanoes that had once erupted above the sea surface and then gradually sank down deeper and deeper below the waves as they aged. From his gravity work, he knew of weird gravity patterns beneath the deep ocean trenches (see chapter 22). From his friends at Lamont, he knew that Bruce Heezen and Marie Tharp had mapped the gigantically long Mid-Atlantic Ridge with its huge rift

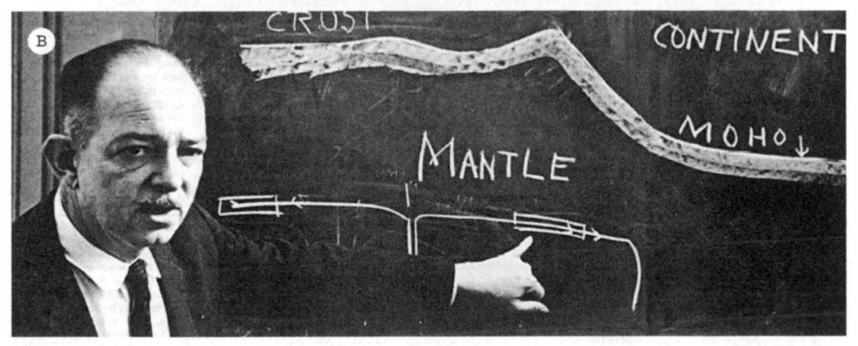

Figure 21.6 ▲

Pioneers of plate tectonics: (*A*) Frederick Vine (*left*) and Drummond Matthews (*right*) at Cambridge in 1963, when they discovered the evidence for seafloor spreading. (*B*) Harry Hess explaining plate tectonics. (Courtesy of Wikimedia Commons)

valley down the middle, proof that the ocean floor was pulling apart in the middle of the ridge (see chapter 22). He was also familiar with Arthur Holmes's suggestion of great convection cells in the mantle that push the crust of the earth along (see chapter 8).

Putting it all together, Hess published a seminal paper in 1962 that proposed almost all the pieces of the modern plate tectonic model, from seafloor spreading on the mid-ocean ridges to the gradual sinking of oceanic crust as it moved away from the ridges. He also recognized that crustal plates descend into the trenches at the leading edge of the moving plate (the yet to be named "subduction zone"). But there was still no hard data to demonstrate that the seafloor actually spread apart and moved like a pair of facing conveyer belts going in opposite directions. Lacking that evidence, he called the paper an "exercise in geopoetry," defusing its bold claims and labeling it as more speculative than it really was.

THE ROSETTA STONE OF THE SEAFLOOR

When most people hear the phrase "Rosetta Stone," they think of a widely advertised language-training program. The actual Rosetta Stone is a huge slab of black granodiorite (figure 21.7) that was discovered in the Nile delta in 1799 by a soldier during Napoleon's conquest of Egypt. The British beat Napoleon in 1801 and captured the Rosetta Stone, and it has been on display in the British Museum since 1802. It is often hard to see because of the mob of people clustered in front of the most famous object in the museum.

The Rosetta Stone is so famous because it has three versions of the same message in three different languages: Egyptian hieroglyphics at the top, Demotic (a cursive form of the language of Egypt) in the middle, and Classical Greek at the bottom (which many scholars can read, including me). Up to that time, no one could translate hieroglyphics, and most Egyptian writing (and thus Egyptian history) was a mystery. But having the same message in a known language (Greek) and an unknown language (hieroglyphics) gave scholars the key to deciphering the unknown language. The stone was finally deciphered by French archeologist Jean-François Champollion in 1822. When he did so, he found the key to the hieroglyphics and, by extension, to ancient Egyptian history. Today, the phrase "Rosetta Stone" is a metaphor for any discovery that provides the key to unlocking a great mystery or opens up a new field of knowledge.

Jumping back to 1963, remember that Vine and Matthews were reading Hess's 1962 paper as they tirelessly reviewed their magnetic data for the Mid-Atlantic Ridge. They knew the mid-ocean ridge was always made of very young lavas that had erupted underwater. They looked closer and noticed the central part of the ridge was always normally magnetized, like the lavas erupting now are, but that there was a symmetrical zone of negative magnetic anomalies on each side (figures 21.5 and 21.8). In fact, a plot of the magnetic stripes from one side of the ridge was the mirror image of the opposite side.

Then Vine and Matthews recalled the recently published work by Cox, Doell, and Dalrymple, laying out the history of the earth's magnetic field across the last 5 million years (figure 21.4B). Then they had a brilliant

Figure 21.7 ▲

The Rosetta Stone, a slab of black granodiorite found near Alexandria, Egypt. It has three different translations of the same message in hieroglyphics (*top*), Demotic (*middle*), and Greek (*bottom*). This gave scholars the key to reading hieroglyphics, which in turn opened the window to understanding ancient Egyptian history. (Courtesy of Wikimedia Commons)

insight. What if the reason for the strong positive anomaly in the center of the ridge was due to the rocks being magnetized in a normal magnetic field, making the overall magnetic signal parallel to the modern field and adding to it, thus giving stronger than average magnetic values? And what if the

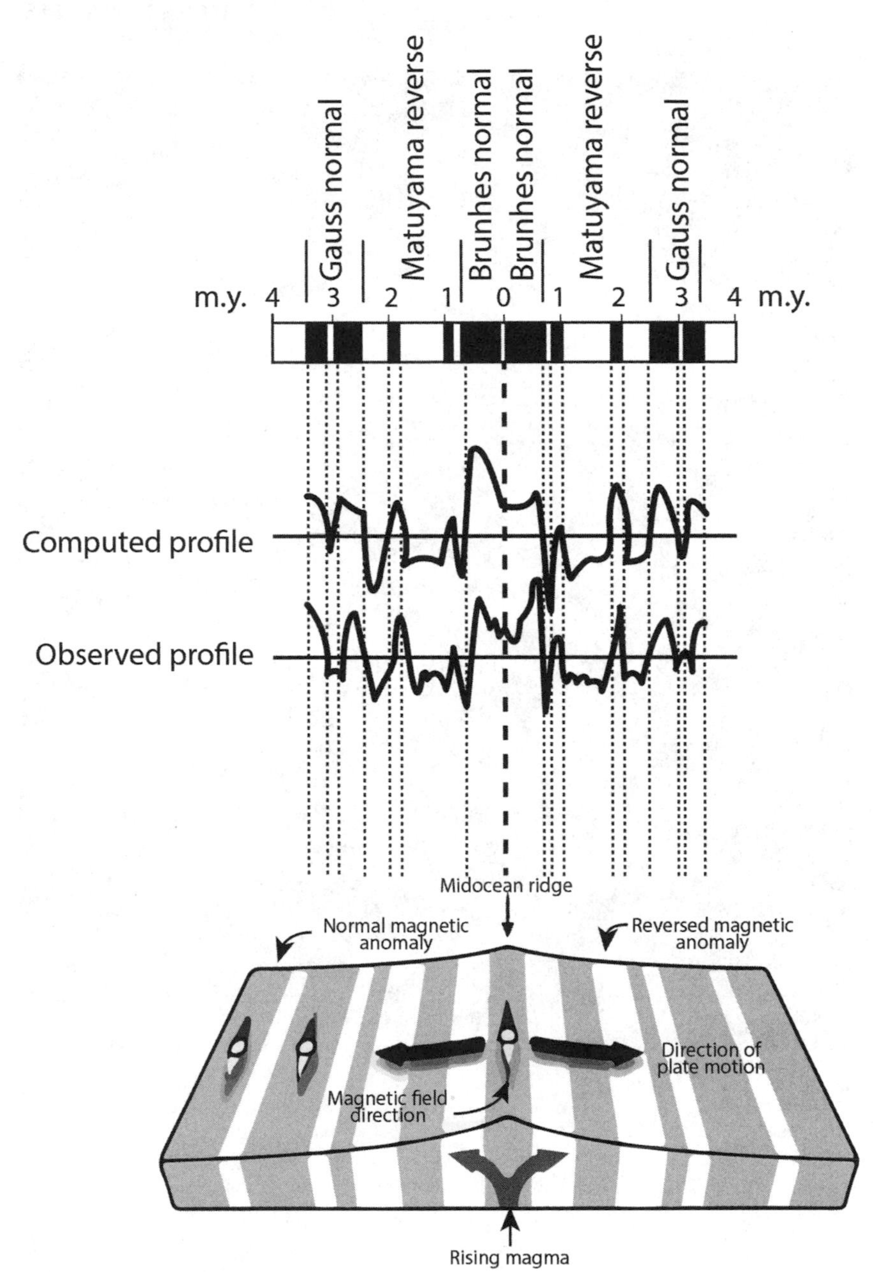

Figure 21.8 ▲

This interpretation of the symmetrical magnetic record (the fluctuating anomalies of magnetic intensity recorded by a shipboard magnetometer on the top) on the mid-ocean ridges argues that they were caused by seafloor spreading (*bottom*), recording the earth's polarity as it erupts in the mid-ocean ridge, then passively moving away like a conveyer belt. (Redrawn from several sources) symmetrical negative anomalies were due to ocean floor that was reversely magnetized back when the field itself was reversed? The reversed direction of the ocean floor lavas of that time would partially cancel the background field felt by the magnetometer and give a weaker than average magnetic reading, or a negative gravity anomaly.

Eureka! It all fit together. The lavas that had erupted underwater, the rift valley in the center, the symmetrical pattern of magnetic stripes being a reflection of symmetrical bands of normal and reversely magnetized rock on the seafloor—all of it fit Hess's seafloor-spreading idea. New ocean floor is constantly being created as it erupts in the rift valley cracks in the center of the ridge as it pulls apart (figures 2.5 and 21.8). As the seafloor lavas cool, they lock in the magnetic field of the earth at the time they form. Then they pull apart, forming new crust in the middle, which would be magnetized in a different direction if the field had flipped. Meanwhile, old crust moves away from the ridge like a pair of conveyer belts. Like an old-fashioned reel of magnetic recording tapes, the "belts" have recorded the signal from the earth's flip-flops of normal and reversed polarity directions as they spooled away from the recording head. Vine and Matthews wrote up their ideas and published them in a landmark paper in 1963, and the plate tectonic revolution was born. (In a strange twist of fate, Canadian geophysicist Lawrence Morley came up with the same idea at the same time, but the journal he sent it to rejected the idea. Most people have never heard of him, although historians of geology now refer to it as the "Vine-Matthews-Morley hypothesis.")

This interpretation of the magnetic anomalies and seafloor spreading would not have been possible without the lucky accident that Cox, Doell, and Dalrymple had independently been documenting the changes in the earth's magnetic field by sampling lavas on land. The magnetic time scale was the Rosetta Stone that unlocked the secret of seafloor spreading. From that point on, the scientific revolution of plate tectonics swept through the geosciences, changing them forever.

FOR FURTHER READING

Butler, Robert F. *Paleomagnetism: Magnetic Domains to Geologic Terranes*. London: Blackwell, 1991.

Cox, Allan, ed. *Plate Tectonics and Geomagnetic Reversals*. San Francisco: Freeman, 1973.

Cox, Allan, and R. B. Hart. *Plate Tectonics: How It Works*. New York: Wiley-Blackwell, 1986.

McElhinny, Michael W., and Phillip L. McFadden. *Paleomagnetism: Continents and Oceans*. New York: Academic, 2000.

Molnar, Peter. *Plate Tectonics: A Very Short Introduction*. New York: Oxford University Press, 2015.

Oreskes, Naomi. *Plate Tectonics: An Insider's History of the Modern Theory of the Earth*. New York: Westview, 2003.

Tauxe, Lisa. *Essentials of Paleomagnetism*. Berkeley: University of California Press, 2010.

22 THE PUZZLE OF SUBDUCTION ZONES

BLUESCHISTS

The Franciscan mélange contains rock of such widespread provenance that it is quite literally a collection from the entire Pacific basin, or even half of the surface of the planet. As fossils and paleomagnetism indicate, there are sediments from continents (sandstones and so forth) and rocks from scattered marine sources (cherts, graywackes, serpentines, gabbros, pillow lavas, and other volcanics) assembled at random in the matrix clay. Caught between the plates in the subduction zone, many of these things were taken down sixty-five thousand to a hundred thousand feet and spit back up as blueschist. This dense, heavy blue-gray rock, characteristic of subduction zones wherever found, is raspberried with garnets.

—JOHN MCPHEE, *ASSEMBLING CALIFORNIA*

PUZZLE #1: VOYAGE TO THE BOTTOM OF THE SEA

Until my generation, one of the great unknowns of science was the shape of the seafloor. Even though it makes up 71 percent of the earth's surface, very little was known of it before the 1950s. This all changed when pioneering geologist Marie Tharp and her partner (both research and personal) Bruce Heezen (figure 22.1) collated all the data from echo sounders collected by ships from Lamont-Doherty Geological Observatory (now Lamont-Doherty Earth Observatory) since the late 1940s and began to plot these depth profiles on a map. The collection of oceanographic data was Heezen's job, because back in the 1950s women were not allowed on those ships. Nevertheless, Tharp was up for her part of the task. She had several college degrees, specializing in mathematics, geology, cartography, and other

Figure 22.1 ▲
Marie Tharp and Bruce Heezen pointing at their map of the seafloor in the early 1960s. (Courtesy of Wikimedia Commons)

sciences, and refused to be remanded to secretarial duties at Lamont. Instead, she worked tirelessly, converting all the depth data from mountains of strip charts produced by the PDR (precision depth recorder) into transects across a map and then filling in the gaps using her knowledge of geology to make a complete physiographic map of the sea bottom.

By the early 1950s, she had plotted Heezen's data and made a complete map of the Atlantic Ocean, the first ever developed, published in 1957. But even as early as 1952–1953, she was the first person to realize that there were mid-ocean ridges spanning the whole globe and that they were the longest mountain ranges on earth. Even more importantly, she saw that there was a huge rift valley larger than the Grand Canyon down the entire axis of the ridge. As a geologist, she knew that rift valleys meant the crust was pulling apart, so she was convinced that the seafloor must be spreading. But Heezen and their boss, Maurice "Doc" Ewing, who founded and ran Lamont, were not so sure. They were the senior authors on the map and papers that came out, and her ideas were suppressed or ignored. A few years later, people such as Harry Hess and Frederick Vine and Drummond Matthews (see chapter 21) connected all the pieces and got the credit for the discovery and confirmation of seafloor spreading. But Marie Tharp was the first to discover the evidence and make the right interpretation, 10 years before Vine and Matthews published the proof.

Tharp and Heezen finished mapping the entire ocean floor by the late 1960s, resulting in the iconic map of the ocean bottom published in 1977 by National Geographic that everyone has grown up with since. But when I was in school, globes and maps only showed the world's oceans as solid pale blue, because almost nothing was known about what lay beneath the waves. I met both Heezen and Tharp when I was a student at Lamont in 1976, and I remember the shock everyone at Lamont felt when we heard the news that Heezen had died of a heart attack in 1977 while on a submarine mapping and photographing the seafloor. Tharp finished her mapping and retired from Lamont in 1983, and she lived comfortably on the income from her maps until she passed away in 2006 at age 86. She mapped more of the earth's surface than any person ever did or ever will, yet very few people have even heard of her, or how she was such a pioneer for women in a science that was dominated by men all her life.

One of the great puzzles of the oceans showed up strikingly on Marie's maps: the enormously deep regions of the seafloor known as oceanic trenches. There are about 50 of them around the world's ocean, stretching 50,000 kilometers (30,000 miles), although they are so narrow they comprise only 0.5 percent of the sea bottom (figure 22.2). Almost all of them are on the rim of the Pacific Ocean, although a few are on the Indian Ocean side of Indonesia and elsewhere. Most are about 3,000 to 4,000 meters (2.0–2.5 miles) deep, although the Marianas Trench is 11,034 meters (36,201 feet, or almost 7 miles) deep, by far the deepest feature on the earth's surface. In fact, it is so deep that you could stick Mount Everest in it and still have 2,186 meters (7,173 feet) to spare. A few of these trenches were detected by the pioneering research ship HMS *Challenger*, a sailing ship that traveled all over the world's oceans between 1872 and 1876, getting depth measurements and dredging up samples. They discovered what is now known as the Challenger Deep, one of the deepest spots on the Marianas Trench.

At that depth, the conditions are so cold and dark and the water pressure so intense that very few organisms can survive, and even most submarines cannot stand the pressure. Finally, in 1960 a specially designed chamber, the bathyscaphe *Trieste* (figure 22.3), which had a thick shell

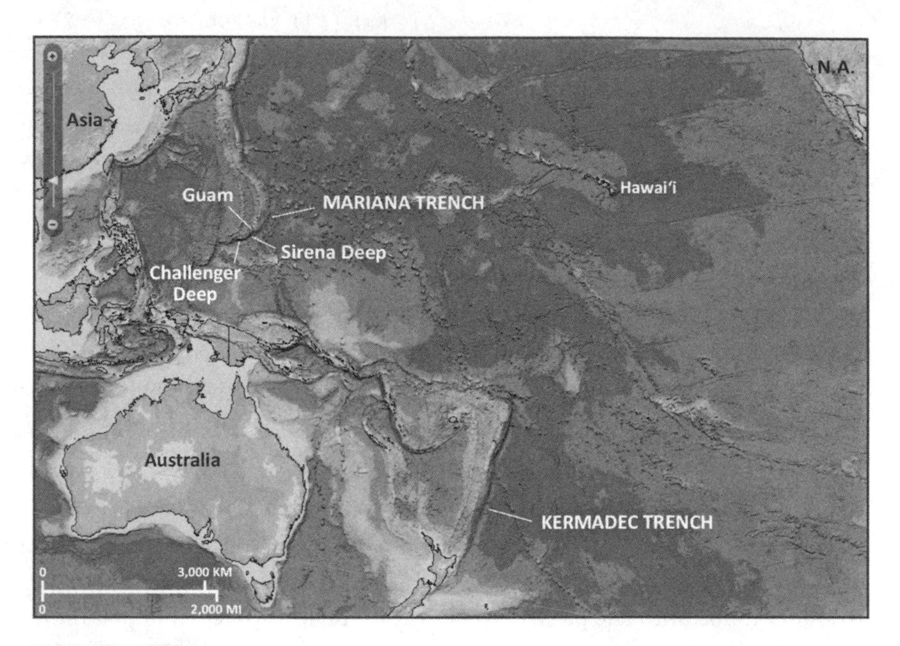

Figure 22.2 ▲

Map of the major oceanic trenches in the western Pacific Ocean. (Courtesy of the U.S. National Oceanic and Atmospheric Administration)

Figure 22.3 ▲

The bathyscaphe *Trieste*, the first vessel to dive to the bottom of the Marianas Trench. (Courtesy of Wikimedia Commons) specially constructed to avoid being crushed by the water pressure, was lowered down to the bottom of the Marianas Trench. It was the first to collect samples and take photographs of the deepest parts of the ocean. Since that time, a number of specially designed human-driven submarines and robotic submersibles have been down to the bottom of the Challenger Deep and other trenches. One of the most recent descents was by filmmaker James Cameron (who directed *Titanic*, *Avatar*, and the *Terminator* movies), who piloted a special submarine known as the *Deepsea Challenger*. On March 26, 2012, Cameron took almost 3 hours to descend to 10,898 meters (35,756 feet), then spent 3 hours exploring before returning to the surface in 2 hours. It was only the second such dive into the deepest part of the ocean, and the first to spend such a long time exploring, and Cameron still holds the record for the deepest manned descent.

Even though Tharp and Heezen had documented many such deep-sea trenches, the origin of these trenches was still a mystery in the 1950s. One key development was the gravity survey of the ocean floor conducted in a submarine by Dutch geophysicist Felix Andries Vening Meinesz. Starting in 1923 and continuing until 1939, he crammed his 6-foot-plus frame into a tiny sub and sailed across the oceans of the world, year after year. Then World War II began, the Germans overran neutral Netherlands, and Vening Meinesz had to put his science aside, fighting in the Dutch Resistance instead. After the war, he resumed his duties at the University of Utrecht. By 1948 he had published his decades of research on the seafloor gravity profiles. One of the most striking results was the extremely low gravity beneath the oceanic trenches. Using the basic principles of gravimetry, this indicated that the crust beneath the trenches was much less dense than expected. By contrast, the density of the seafloor in most places was very high because of the rocks of the very dense mantle that lie just below the thin oceanic crust. In other words, something like low-density crustal material must have displaced the dense mantle at great depth beneath the trench.

PUZZLE #2: ZONE OF PLUNGING EARTHQUAKES

As the postwar boom in earth sciences was getting underway, seismology was making progress by leaps and bounds. Flush with funds from the efforts to monitor Soviet nuclear tests (nuclear explosions create distinctive shock waves that seismographs can pick up), seismologists developed more and more sophisticated ways to interpret the waves coming through the earth. They had already used the behavior of seismic waves to figure out the structure and depth of the mantle and core and to determine that the outer core was fluid. Now they found that seismic waves could tell them even more.

Working independently in the late 1920s and 1930s, Caltech seismologist Hugo Benioff and Japanese seismologist Kiyoo Wadati both developed methods to find out how deep below the surface any earthquake had occurred. Then they both began to plot the many earthquakes that were occurring near the oceanic trenches. To their surprise, there was a clear-cut pattern (figure 22.4). The shallow earthquakes were just beneath the oceanic trench and the edge of the continent. But the deeper the quakes were, the farther

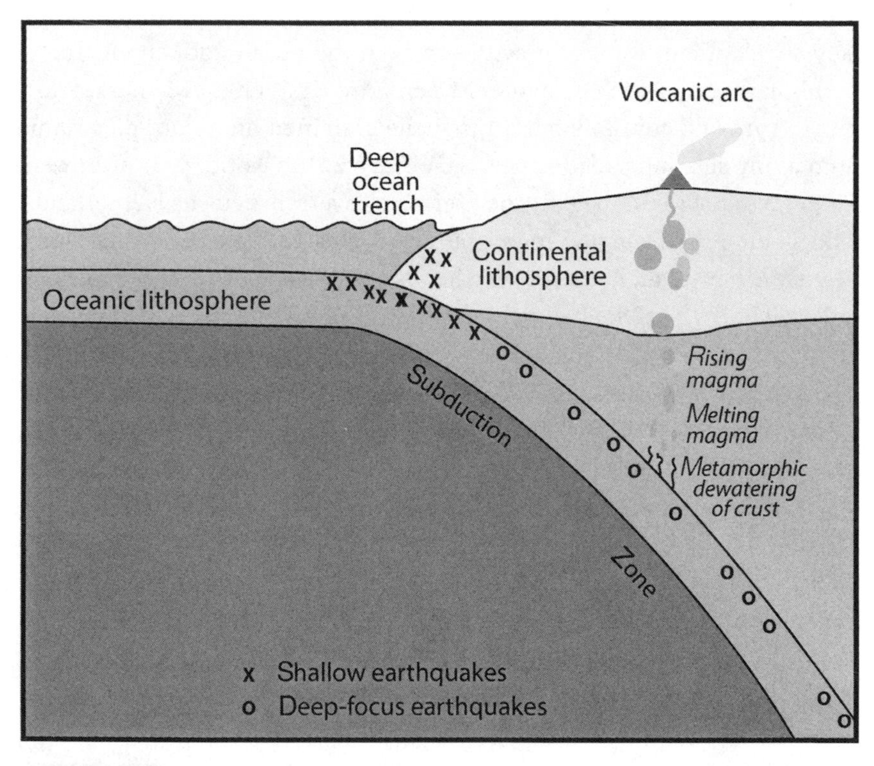

Figure 22.4 ▲

The plunging zone of earthquakes beneath the oceanic trench and the volcanic arc, known as the Wadati-Benioff zone. (Redrawn from several sources)

they were located beneath the continental edge and away from the trench. In fact, they formed a distinctive zone of plunging earthquake occurrences, sometimes reaching hundreds of kilometers below the surface.

What could this pattern mean? Benioff finally published his results in 1949; Wadati published the same basic idea in 1928, but few people knew about it. They both pointed out that the pattern was found in nearly every oceanic trench around the world. But they had no explanation. It was like having a piece of a jigsaw puzzle with no other pieces. You have no idea what it means until you see the rest of the puzzle.

About the only explanations published for these strange features of the oceanic trenches (gravity profile, then eventually seismicity) were a 1938 paper by Harry Hess and a 1939 paper by David T. Griggs, a geophysicist who helped found the RAND Corporation. As Air Force chief scientist, Griggs helped supervise the development of the hydrogen bomb in 1951. In their papers, Hess and Griggs each proposed that thermal convection in the mantle caused the oceanic crust to become downbuckled and folded in upon itself, like a crumpled carpet, explaining the unusual thickness of low-density crustal rock beneath the trenches (figure 22.5). Hess called this odd structure a "tectogene." The idea was ahead of its

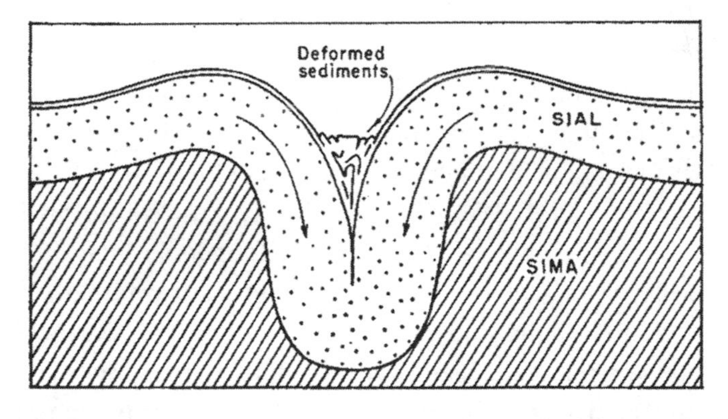

Figure 22.5 ▲
Harry Hess's 1938 concept of a "tectogene," a buckling of the earth's crust beneath the oceanic trench, which was imagined to explain the low gravity and thick crustal rocks in the region and the plunging zone of seismicity documented by Wadati and Benioff. The "SIAL" is of crustal rocks made of silicon and aluminum; the "SIMA" is made up of mantle rocks made of silicon and magnesium. (Modified from H. H. Hess, "Gravity Anomalies and Island Arc Structure with Particular Reference to the West Indies," *Proceedings of the American Philosophical Society*, 79 (1938):71–96.)

time, but eventually would become obsolete when plate tectonics came along (although I still remember seeing it in some of my college textbooks that were published in the 1960s, before plate tectonics forced everyone to rewrite the textbooks). Benioff thought there might be something like a tectogene crustal fold structure causing the plunging zone of earthquakes, but almost everyone was stumped as to their real cause. It would be almost 20 years later before the Wadati-Benioff zones would help decipher an even bigger puzzle.

PUZZLE #3: COOL UNDER PRESSURE

For many years, geologists had been mapping the rocks of the Coast Ranges of California. In many places, they found a peculiar blue-colored metamorphic rock they called "blueschist" (figure 22.6). It has the same strongly planar foliation found in other kinds of schists, but it is rare in most parts of the world, and at first was only known from the California Coast Ranges and Japan. Geologists looked at 30-micron-thick samples of it under the microscope and found it was made of peculiar minerals. One was a unique, bluish, sodium-rich amphibole known as glaucophane (whose name means "appears blue-green" in Greek). Although first described from the Greek Cycladic Islands of the Aegean Sea, glaucophane soon became much better known in California. These rocks had been informally called "blueschist" by many geologists, but no one actually published the name until 1962, when U.S. Geological Survey scientist H. Bailey did so in an abstract.

Another unusual blue mineral was discovered in 1895 in blueschists of the Tiburon Peninsula of Marin County, California. It was formally named lawsonite, after Andrew Lawson, one of the founders of the University of California–Berkeley geology program (and the man who named the San Andreas fault and led the study of the 1906 San Francisco earthquake—see chapter 23). The mineral was discovered by two of his students, Charles Palache and Frederick L. Ransome, who named it in his honor. Geologists had soon accumulated a long list of the odd minerals that occurred in blueschists, but it was still a complete mystery how they had formed. There were also other minerals, such as the aluminosilicate kyanite, which is steely blue in color, and the pyroxene known as jadeite, one of the two minerals that produce jade gems.

Figure 22.6 ▲
A classic blueschist, made of highly foliated glaucophane, lawsonite, and other blue minerals. (Courtesy of C.-T. Lee)

It was not until the 1940s and 1950s that geologists were able to synthesize some of the key minerals in metamorphic rocks. Using extremely high-temperature, high-pressure lab furnaces, they could take samples of rocks and heat them and put them under huge pressure, then see what minerals were produced. By this method, they were able to determine that certain metamorphic minerals were like geobarometers and geothermometers. Thanks to the lab data, they were known to form only under a certain range of temperatures and pressures, so their occurrence in any metamorphic rock told us the range of pressures and temperatures that the rock had experienced.

From this, pioneering scientists, such as the Finnish geologist Pentti Eskola in 1939, were able to determine that rocks that had long been called "greenschists" because of their green minerals, like the mica known as chlorite or the green minerals like epidote and the amphiboles tremolite and actinolite, had undergone relatively low pressures and temperatures (2–8 kilobars of pressure and only 330–500°C). If a rock is pushed farther down into the earth's crust, it reaches a region of intermediate pressures and temperatures (4–12 kilobars and 500–700°C). The geologists' term for intermediate-grade metamorphism is "amphibolite," because these rocks tend to be rich in the black amphibole hornblende. Finally, a rock can descend to the very deepest part of the continental crust, where there are extreme pressures (6–14 kilobars) and temperatures (at least 700°C) that create the highest-grade metamorphics. These rocks have a gneissic texture and are sometimes called "granulites." All three of these metamorphic rock types are produced by regional metamorphism, such as when a collision between continents creates a huge uplifted mountain belt (such as the Himalayas today). Rocks that undergo shallow burial will only turn into greenschists, while those buried more deeply will become ampibolites or, finally, granulite gneisses, if they are many kilometers down into the crust at the root of the mountains.

This was the pattern of the majority of the metamorphic rocks in the world, and the distribution of these rocks was well mapped and understood. But what about blueschists? Why did they have these peculiar minerals? Why were they so rare and only found in a few places: the Greek isles, California Coast Ranges, coastal Japan, and a few other spots? This remained a mystery for many years. Eskola was puzzled about them in 1939, although he suspected they were high-pressure rocks, because they looked to be intermediate to known high-pressure rocks such as eclogites (which we now know come from near the mantle). In the late 1950s, a number of geologists thought that the jadeite and quartz assemblage of some blueschists were the high-pressure analogue of greenschists, which were well understood.

Finally, in the early 1960s, Stanford geologist W. Gary Ernst and other scientists did the crucial lab experiments to find out just what kind of conditions produced rare minerals such as glaucophane and lawsonite and the blueschist assemblage in general. After many years of lab experiments, the answer emerged, but it was just as puzzling as before: extremely high pressure (more than 4–6 kilobars) but relatively low temperatures (below 400°C). This was peculiar. Normally, rocks that get buried deep enough to experience such high pressures are also cooked to high temperatures. How could you get a rock that was "cool under pressure" and apparently never cooked to a high temperature despite experiencing high pressure? And how did a rock go from deep in the crust where such high pressures reign and then back up to the surface? And why were they found only in the California Coast Ranges and a few other places?

Another piece of the puzzle had been added. But there were far too many missing pieces to put it together and make sense of it yet.

PUZZLE #4: CHAOTICALLY JUMBLED ROCKS

For decades, geologists had mapped many of the weird rocks of the California Coast Ranges from San Francisco south to San Luis Obispo as the "Franciscan Formation" (a term coined by Andrew Lawson himself). However, these rocks did not show simple layer-cake stratigraphic relations or much continuity over distance like a true sedimentary formation. Instead, the "Franciscan" rocks appeared to be highly sheared, deformed, and "sliced and diced," as if they had been chopped up in a blender. Because of this appearance, they were given the name "mélange," from the French word for "mixture" (figure 22.7). Typical rocks included lots of deep-sea shales and chert layers but also turbidite sandstones (chapter 15), although these sedimentary units had all been sliced up and deformed, so they could not be traced any significant distance. The bedrock often would change abruptly as you crossed from one fault block to the next. There were lots of peculiar rocks as well. Ophiolites (chapter 2) crop out in many places in the Coast Ranges, but they were not recognized as slivers of oceanic crust until the 1970s. These rocks were metamorphosed from basalt into a fibrous greenish-black mineral known as serpentine (from its smooth snakeskin texture). This produced large quantities of the rock made of the mineral serpentine (and minerals such as asbestos), known as serpentinite. In

Figure 22.7 ▲

Typical outcrop of mélange, highly shredded and discontinuous, here in the beach outcrops below San Simeon and Hearst Castle, California. (Photo by the author)

addition, serpentines weathered to soils rich in magnesium that would only support certain kinds of plants. Even more peculiar were the large chunks of blueschist, which were not fully understood until the 1960s as products of high pressures but relatively low temperatures.

How did rocks get all sheared and shredded like this? Why were these mélanges found only in places like the California Coast Ranges? And why were they composed entirely of a weird mix of deep-sea sediments, ophiolites, and blueschists—all rocks that were extremely rare in any other setting? The puzzle remained unsolved for most of the early and mid-twentieth century. Lots of pieces of the puzzle were coming together by the early 1960s, but no one could see the big picture yet.

SOLUTION #1: SUBDUCTION LEADS TO OROGENY

Running through all the early ideas about mantle convection and the possible drift of continents there seemed to be a suggestion that the plates produced where magma comes up from the surface must somehow sink down into the mantle where the currents in the upper mantle also plunge down into the mantle. This is implied in diagrams (figure 8.2) from Arthur Holmes's pioneering geology textbook of 1944. As Holmes wrote:

> They must founder into the depths, since there can be nowhere else for them to go. Now this is precisely what would be most likely to happen when two opposing currents come together and turn downwards beneath a cover of basaltic composition. On the ocean floor the expression of such a down-turning of the basaltic layer would be an oceanic deep. The great deeps bordering the island festoons of Asia and the Australasian arc (Tonga and Kermadec) probably represent the case where the sialic edge of the continent has turned down to form the inner flanks of a root, while the oceanic floor contributes to an outer flank . . .
>
> During large-scale convective circulation the basaltic layer becomes a kind of endless traveling belt on top of which a continent can be carried along, until it comes to rest (relative to the belt) when its advancing front reaches the place where the belt turns downwards and disappears into the earth.

Given that this was written in 1944, long before most of the evidence for plate tectonics came along, it sounds surprisingly modern. Holmes not only had anticipated seafloor spreading, the "endless traveling belt" of basaltic oceanic crust rising from the seafloor spreading zone, but especially how it went plunging back down into the mantle—a concept that is basically subduction but was not yet named that way.

Even as early as 1939, David Griggs was invoking the idea of plunging zones of earthquakes beneath oceanic trenches as possible evidence of the underthrusting of one plate into another along convection currents. In his words, leading seismologists "all agree that foci of deep earthquakes in the circum-Pacific region seem to be on planes inclined about 45° toward the continents. It might be possible that these quakes were caused by slipping along convection-current surfaces."

Harry Hess, in his famous 1962 "geopoetry" paper (see chapter 21), spelled it out clearly. He viewed the ocean floor as an ephemeral feature, borne along by the convection currents in the mantle, created at the ridges, and sinking back down where the convection currents went back down to the mantle. In his words:

> The leading edges are strongly deformed when they impinge upon the downward moving limbs of the convecting mantle. The oceanic crust, buckling down into the descending limb, is heated and loses it water to the ocean. The cover of oceanic sediments and the volcanic seamounts also ride down into the jaw crusher of the descending limb, are metamorphosed, and eventually probably welded onto continents.

Here we see nearly all the elements of modern thinking about subduction zones: the downgoing slab, the heating and losing water (which mostly stimulates melting of magmas to form arc volcanoes), the pieces of oceanic crust, seamounts, and other materials getting metamorphosed and deformed in the "jaw crusher of the descending limb," and the eventual welding onto the continents.

Once Vine and Matthews published the evidence for seafloor spreading in 1963, an avalanche of groundbreaking papers by J. Tuzo Wilson, Dan McKenzie, W. Jason Morgan, Xavier Le Pichon, and many others came out in the mid-1960s. Nearly all of these talked about "trenches" and "compressive boundaries," so the major mysteries of what formed oceanic trenches, and why their gravity showed less dense rock, and why they had the Wadati-Benioff zones of plunging earthquakes had all been connected to the emerging models of plate tectonics.

PUZZLE #5: THE ALASKAN EARTHQUAKE

But a key event that literally shook the world, and showed subduction in action, was the Alaskan earthquake of 1964. Striking at 5:36 in the afternoon on Good Friday, March 27, 1964, it was the most powerful earthquake ever recorded in the United States, with a moment magnitude of 9.2. Alaska had only been a state for 5 years at that point. Some towns in southern Alaska were torn apart, and some areas just sank into the ocean and vanished. Huge tsunamis swept away everything in the coastal zone, while other areas turned into quicksand and sank, or huge landslides caused houses to ride down the slipping ground like roller-coaster cars. Despite the huge magnitude of the quake and the 7 minutes of intense shaking in some places, it is amazing that only 136 people died. This was a bit of luck, because most people were already safe at home after work for the Good Friday holiday, and also because the population of Alaska was so small back then.

Although many people dealt with the logistical problems of the earthquake emergency, geologists were dispatched to study the quake in great detail and survey its effects. One of these was U.S. Geological Survey geologist George Plafker, who was on the scene when the quake happened. Seismologists soon established that it was a huge thrust fault, with a portion of the Pacific seafloor being pushed beneath the Aleutian Trench and Alaska to the north. Plafker found a very interesting pattern in the crustal rocks above the thrust (figure 22.8A). The rocks immediately to the north of the Aleutian Trench, from Prince William Sound, to the south shore of Kodiak Island, had been uplifted some 9.1 meters (30 feet) in the air. The piers in the region saw the water recede like during a low tide, but the water never came back, and all the marine life that used to be in the tide pools died because it was lifted out of the tidal zone (figure 22.8A and B). But even farther north from the trench, in the region to the northwest of the uplifted area (Kenai Peninsula and Kodiak Island and the shores of Cook Inlet), the ground sank 2.4 meters (8 feet), so the tide came in and flooded the coastal region and never left. Huge forests that used to be above the shoreline were flooded with saltwater and died (figure 22.8A and C).

Plafker put all this information together and correctly realized that these were the crustal consequences of the crumpling of the block overlying the huge thrust fault beneath. The region just over the fault zone in the trench had been buckled up by the pressure, while the crust just behind it was buckled down almost the same amount. The 1964 Alaska earthquake was the first demonstration of subduction in action.

By the late 1960s, seismologists were using a new development of an old method to confirm that plate motions were in the same direction as predicted by plate tectonics. Using a tool called first-motion analysis, they could tell from an array of different seismographs which direction the fault

Figure 22.8 ▲

(*A*) The pattern of uplift and subsidence after the 1964 Alaska quake seemed to be due to the underthrusting along the Aleutian Trench. (*B*) Uplifted beach exposures that exposed marine life to dry conditions after the tides could no longer reach them. (*C*) Other places were flooded by tides after the 1964 Alaska quake and sank permanently beneath the sea. (Courtesy of U.S. Geological Survey)

Figure 22.8 ▲
(*continued*)

had moved and what the angle of the fault plane was. Compiling data from thousands of earthquakes on dozens of different plate boundaries, Lamont seismologists Bryan Isacks, Jack Oliver, and my former seismology professor Lynn Sykes showed in 1967 and 1968 that earthquakes proved that crustal boundaries were moving according to plate tectonic models. All the quakes not only showed that the zone of seismicity was plunging beneath the overlying plate (as cruder data had shown Benioff and Wadati) but that the seismic motion demonstrated overthrust of one plane on top of another. By contrast, the quakes from mid-ocean ridges were pulling apart along vertical planes, as would be predicted from seafloor spreading. This research was considered one of the proofs that clinched the reality of plate tectonics.

Through all this burst of scientific activity, what to call this concept was still in the air. Terms such as "underthrust," "downgoing slab," "overriding plate," "overlapping plate," "zone of convergence," "zone of shortening," "consumption zone," "sink," "trench," "zone of destruction of crust," "Benioff zone," "island arc," "down-dragged crust," "downwelling site," "descending limb," "take-up zone," "undercurrent," or "underflow" had all been used in one way or another, with all sorts of misleading connotations. Some, like "Benioff zone," were seismic concepts, while others were structural or tectonic in meaning.

In early 1969, Bill Dickinson of Stanford University organized the first-ever Penrose Conference of the Geological Society of America (GSA). Penrose Conferences are very different from normal GSA meetings, where 5000–6000 geologists over 4 days and about 20 concurrent sessions give quick 15-minute talks that may or may not stimulate any questions or discussions or just go to hear others talk. I've been to three Penrose Conferences, two of which I organized, and they are more like a workshop or brainstorming session than a formal meeting. Every participant must be invited for the contribution he or she could make to a broad topic. Everyone presents something, often for as long as the participants want to discuss the topic, so the schedule is free and flexible. Even more importantly, all participants attend every talk, so it's not like a normal GSA meeting, where there are so many talks in so many rooms that you can only attend a few of them. The whole idea is to bring together people of different disciplines who normally never interact to listen to one another and see what they can learn from one another.

The 1969 Penrose Conference was historic, not only in being the first such conference ever to be organized, but also because Dickinson and his co-conveners brought together nearly all the pioneers of plate tectonics in the same place. Scientists who were not on the cutting edge got a complete picture of the newly emerging plate tectonic synthesis. There were also a lot of the older giants of the field, who had enormous backgrounds in some areas of geology but had not yet seen plate tectonics applied to those specialties. The collective brainpower of the meeting was impressive, considering that more than a dozen were or later became members of the National Academy of Sciences.

Dickinson and his co-conveners held the meeting at the Tony Asilomar resort in Pacific Grove on the Monterey Peninsula. Asilomar is famous for cutting-edge meetings such as the TED conferences, weekends focused on meditation and self-help, and many other types of meetings. It is a perfect place to relax and commune with the universe or discuss important ideas sitting in a hot tub with a good glass of wine and spectacular views of the Pacific Ocean and redwood trees all around you. Held on December 15 to 20, 1969, this first Penrose Conference transformed the entire geological community. All the participants found that they each had a key piece of the puzzle of plate tectonics and that all the pieces fit together—and that the completed picture beautifully explained some important geological puzzles that the old guard had wrestled with for decades. When the conference ended, the participants left with their old notions shattered and their heads full of new ideas, and many of them eagerly jumped onto the accelerating bandwagon that was the scientific revolution of plate tectonics.

One of the prized souvenirs was an ordinary dinner plate smuggled out of the Asilomar cafeteria that was "awarded" to Dickinson by the other conference members. The other participants had inscribed the plate in marker with these words: "Hero of Plate Tectonics Award," "Penrose Conference Asilomar December 1969," and on the rim was written "In Subduction We Trust." As Dickinson himself later commented, "The plate tectonic revolution in geoscience came along just about the time I had my professorial feet solidly on the ground, and I rode that rocket for a number of years. . . . The plate revolution was indeed as much fun as a barrel of monkeys. We kept asking ourselves how we could have been so dumb for so long. But we also knew we were in the catbird seat and could carry the day just by plowing straight ahead."

During the Asilomar Penrose Conference, the participants debated all the confusion of terms about what to call this mysterious region with underthrusting and the plunging earthquakes of the Wadati-Benioff zone. After doing some research into historic precedents, they agreed that the old term "subduction," coined by André Amstutz in 1951 to describe this phenomenon in the Alps, was one of the first published names that could be applied to the concept. Thus, "subduction" was formally defined as the process when one plate plunges beneath another, and the geologic community had a clear terminology for the components of a "subduction zone."

SOLUTION #2: THE SUBDUCTION ZONE COMPLEX

By the late 1960s, the nature of oceanic trenches and crustal plates sliding one beneath another had been well defined by geophysical data, especially from the areas of seismology, gravity, and paleomagnetism. The effects of an actual earthquake caused by one plate sliding beneath another had been documented by the 1964 Good Friday Alaska quake. But what about the peculiar rocks known as mélange or the strange metamorphics called blueschist?

All through the 1960s, marine geology had continued to make big strides. Geologists had carefully surveyed and sampled not only the rocks in the deep ocean trenches but also on each side of the trenches. On the edge of the plates overlying the trenches, they found peculiar ridges of material that sometimes was so thick it rose above sea level and formed a chain of islands just landward of a trench. Once ships had drilled into these ridges and run seismic reflection profiles through them, the geologists could see that the ridge material had an unusual structure. Drill cores showed that it was composed of highly sheared rock, often with no pervasive bedding. Even more unusual, the oldest slices were on top, and the rocks got younger toward the bottom, the exact opposite of what happened when sedimentary layers slid down one on top of another, with younger rocks on top of older rocks.

This was even more apparent in the seismic profiles, which showed that each ridge was a stack of thrust slices, one on top of another, often intensely folded upon themselves (figure 22.9). In some cases, there were dozens of these slices stacked on their sides, like a deck of cards on its side. Even more revealing, the rocks obtained from drill cores were mostly deformed marine sediments, such as deep-marine shales and cherts and turbidites, but occasionally the researchers found slices of ophiolite and other peculiar rocks.

All of this marine geological research led researchers to realize that this peculiar pile of rocks was material scraped off the downgoing slab as it plunged into the trench and down into the mantle. Its lack of continuity and intense deformation made sense if it was being intensely sheared and sliced and crumpled by the process of scraping between two plates. It is analogous to rocks being scraped off by a bulldozer blade, with the first rocks being pushed to the top of the stack as the blade slices off new rocks and slides them beneath the old ones. The longer subduction occurred, the larger and taller the stack of trench scrapings would be.

Almost as soon as these observations emerged from marine geology, geologists on land realized that this was exactly the process that could explain the peculiarities of mélange. In a series of classic papers published between 1969 and 1971, Gary Ernst and Bill Dickinson (both at Stanford then) and Ken Hsü (then of the U.S. Geological Survey) made the connection and argued that the chaotically deformed "Franciscan Formation" was not a conventional sedimentary formation at all, but a tectonic assemblage—in other words, an accretionary wedge complex. All the pieces of that puzzle made sense.

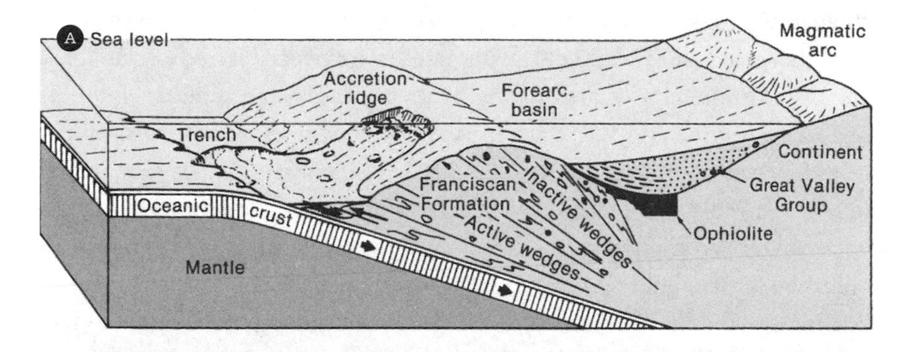

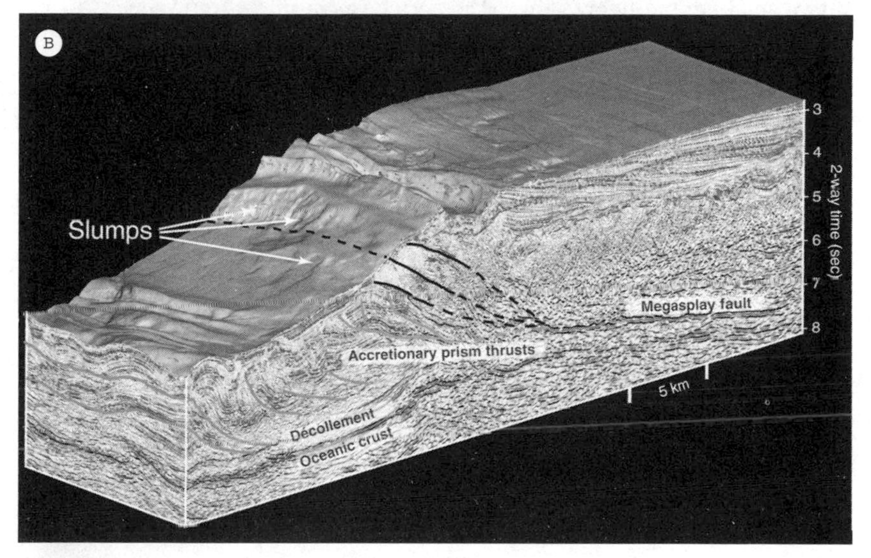

Figure 22.9 ▲

(*A*) Diagram of the accretionary prism in front of the volcanic arc. It is built of material scraped form the downgoing slab and plastered onto the base of the overlying plate. (*B*) Seismic profile through an accretionary prism, showing the intense faulting and folding within it. ([*A*] Modified from Donald R. Prothero and Robert H. Dott Jr., *Evolution of the Earth*, 8th ed. [New York: McGraw-Hill, 2010]; [*B*] Courtesy of Wikimedia Commons)

Only one more piece of the puzzle remained: the peculiar blueschists. In the early 1960s, Ernst and others had shown that they formed under extremely high-pressure but low-temperature settings. Where could such places exist? The fact that blueschists were found in mélanges and in association with ancient volcanic arcs suggested that they too were the products of subduction. In a series of papers from 1970 to 1973, Ernst and others showed that subduction zones might be the ideal place for such peculiar metamorphic conditions. The downgoing slab of oceanic crust was not only cold after millions of years of slowly moving across the seafloor, but also deeply weathered and full of water. When it plunged down into the mantle, it brought cold wet rock into a region of very high pressure. It took a long time for the cold slab to heat up to the temperature of normal deep crustal–upper mantle rocks. Such conditions would be the ideal setting to make blueschists. The cold slab would not only remain "cool under pressure," but some of it might be pushed up into the underplating of the accretionary prism and eventually brought to the surface as more tectonic slices pushed the rocks even farther up into the air.

Today, we can find examples of ancient subduction zones, with their accretionary wedges, blueschists, mélanges, and ophiolites, in many parts of the world. But connecting all the pieces of this puzzle was a slow process that only accelerated with the beginning of marine geology and then with the birth of plate tectonics.

FOR FURTHER READING

Cox, Allan, ed. *Plate Tectonics and Geomagnetic Reversals.* San Francisco: Freeman, 1973.

Cox, Allan, and R. B. Hart. *Plate Tectonics: How It Works.* New York: Wiley-Blackwell, 1986.

Felt, Hali. *Soundings: The Story of the Remarkable Woman Who Mapped the Ocean Floor.* New York: Holt, 2013.

Molnar, Peter. *Plate Tectonics: A Very Short Introduction.* New York: Oxford University Press, 2015.

Oreskes, Naomi. *Plate Tectonics: An Insider's History of the Modern Theory of the Earth.* New York: Westview, 2003.

23 | EARTHQUAKE! THE SAN ANDREAS FAULT

TRANSFORM FAULTS

In clear weather, a pilot with no radio and no instrumentation could easily fly those four hundred miles navigating only by the fault. The trace disappears here and again under wooded highlands, yet the San Andreas by and large is not only evident but also something to see—like the beaten track of a great migration, like a surgical scar on a belly. In the south, . . . it cuts through two high roadcuts in which Pliocene sedimentary rocks look like rolled up magazines, representing not one tectonic event but a whole working series of them exposed at the height of the action.

—JOHN MCPHEE, *ASSEMBLING CALIFORNIA*

SAN FRANCISCO, 1906

It's 5:00 A.M. on April 18, 1906, a quiet morning in the City by the Bay. Only a few patrolling policemen and deliverymen are up and moving around, while the rest of the city sleeps. The night before, April 17, had been unusually hot. Many of the wealthy had been to a performance of *Carmen* at the Grand Opera House, starring the legendary tenor Enrico Caruso in the role of Don Jose. (Ironically, the newspapers of the day featured accounts of eruption of Mount Vesuvius in Italy, not far from Caruso's hometown of Naples.)

From a sleepy little Mexican town back in the 1830s and 1840s, San Francisco had grown to a huge city of more than 400,000 residents thanks to the California Gold Rush in 1848 and afterward. As it grew in size, it also grew in hazards. Its ramshackle construction of wood and canvas from the Gold Rush days had burned down at least six times during repeated fires between 1848 and 1851. There were also small earthquakes in 1836 and 1868, but it had been 38 years since the last quake, and most residents had either forgotten it or had arrived after the quake or had not even been born in 1868. By 1906 most of the old wooden buildings had been replaced by shiny newer buildings with steel frames and brick-and-mortar exteriors. Many had large cisterns of water on their roofs to provide pressurized water for firefighters. San Francisco had one of the largest and most professional firefighting systems in America.

At 5:12 A.M. the quiet streets and buildings were suddenly shaken by a series of large jolts. One cop on duty described the motion of the ground like a wave of a rough sea rolling down the street (figure 23.1A). Another policeman, Officer Jesse B. Cook, was standing at the eastern end of Washington Street and was one of the first to witness the waves of energy approaching from the north. There were waves of water advancing down the street, and the entire street was undulating. The buildings and pavements were lifted up and toppled. In his report, he said:

> The earth seemed to rise under me, and at the same time both Davis and Washington streets opened up in several places and water came up out of these cracks. The street seemed to settle under me, and did settle in some places from about one to three feet. The buildings around and about me began to tumble and fall and kept me pretty busy for a while dodging bricks. I saw the top story of the building at the southwest corner of Washington and Davis fall and kill Frank Bodwell.

A night clerk at the Valencia Street Hotel who ran from the building described it this way: "The hotel lurched forward as if the foundation were dragged backward from under it, and crumpled down over Valencia Street. It did not fall to pieces and spray itself all over the place, but telescoped down on itself like a concertina."

The people who were on the first three floors were crushed to death (at least 100 were killed), while those who happened to have rooms on the fourth floor simply stepped out onto the street. Another witness named P. Barrett wrote: "We could not get to our feet. Big buildings were crumbling as one might crush a biscuit in one's hand. Ahead of me a great cornice crushed a man as if he were a maggot."

Figure 23.1 ▲

The San Francisco earthquake of 1906: (*A*) buckled and cracked streets; (*B*) toppled buildings; (*C*) crowds watching the fires traveling up the hill toward them on Golden Gate Heights; (*D*) the conflagration around San Francisco City Hall, with soldiers on guard to prevent looting. (Courtesy of U.S. Geological Survey)

Figure 23.1 ▲
(continued)

The Palace Hotel was the grandest in all of San Francisco, playing host to celebrities like Caruso and also presidents and kings. It was seven stories tall, with over 800 rooms and four of the newly invented elevators, and was the largest hotel in the country at the time. It also had 700,000-gallon water cisterns made of iron that had been built in a space beneath the roof to help the firemen. When the shaking started, the horses in the carriage entrance bolted and the trees swayed, but otherwise the building held up. Caruso had gone to bed only 2 hours earlier, after his post-performance meal, but he was severely shaken and panic-stricken. Different versions of how he reacted to the events have been published, but one account says that he had put on a fur coat over his nightclothes and immediately left town, muttering "'Ell of a town! 'Ell of a town! I never come back!" And he never did.

Policeman Harry Walsh witnessed the death and destruction and saw huge cracks open in the pavement on Fremont Street, which closed and reopened as the shock waves passed. Then he saw a herd of longhorn cattle stampeding toward him along Mission Street from the direction of the docks. Apparently, they had just been unloaded from an inbound ship and were being driven to the stockyards south of town when the quake occurred. The Mexican *vaqueros* had fled in panic, leaving their cattle to run through the streets of the city. As Walsh wrote:

> While a lot of them were running along the sidewalks of Mission Street, between Fremont and First streets, a big warehouse toppled onto the thoroughfare and crushed most of them clean through the pavement into the basement, killing them and burying them outright. The first that I saw of the bunch were caught and crippled by falling cornices, or the like . . . and were in great misery. So I took out my gun and shot them. Then I had only six shots left, and I saw that more cattle were coming along, and that there was going to be big trouble.
>
> At that moment, I ran into John Moller, who owned the saloon . . . I asked him if he had any ammunition in his place, and if so, to let me have some quick. He was very scared and excited over the earthquake and everything; and when he saw the cattle coming along, charging and bellowing, he seemed to lose more nerve.
>
> Anyway, there was no time to think. Two of the steers were charging right at us while I was asking him for help, and he started to run for his saloon. I had to be quick about my part of the job because, with only a revolver as a weapon, I had to wait until the animal was quite close before I dared fire. Otherwise, I would not have killed or even stopped him.
>
> As I shot down one of them I saw the other charging after John Moller, who was then at the door of his saloon and apparently quite safe. But as I was looking at him and the street, Moller turned, and seemed to become paralyzed

> with fear. He held out both hands as if beseeching the beast to go back. But it charged on and ripped him before I could get near enough to fire. When I killed the animal it was too late to save the man . . .
>
> Then a young fellow came running up carrying a rifle and a lot of cartridges. It was an old Springfield and he knew how to use it. He was a cool shot, and he understood cattle, too. He told me he came from Texas . . . we probably killed fifty or sixty.

The actual shaking lasted about 40 seconds, but seemed like an eternity to those who experienced it that morning. Nevertheless, the devastation was total. Nearly every brick building and chimney had fallen down (figure 23.1B), since they were not reinforced with steel rods or framing, as is required in California now. Wood-frame and steel-frame buildings did better at first, but the wooden buildings were still susceptible to the fires that broke out immediately as all the oil lamps fell and hearths with fires in them shattered (figure 23.1C). Soon the fire was raging out of control. It burned for 4 days, destroying about 28,000 buildings, leveling more than 75 percent of the city, and ultimately producing 10 times as much damage as the quake itself.

Despite all their preparations, the fire department was helpless. The fire chief was killed in the initial quake, and most of their water supply was cut off because the shaking had ruptured the water pipes. The cisterns on the tops of buildings helped a bit, but they were not enough to battle the conflagration. Soon, 2000 soldiers from the Presidio (figure 23.1D) roamed the streets with orders to shoot any looters (despite no formal declaration of martial law). The firefighters were so desperate to stop the blaze that they tried dynamiting buildings in its path to create a firebreak. "One of the problems was the type of explosives that they used," according to Philip Fradkin, author of *The Great Firestorms of 1906*. "Gunpowder is flammable and spreads fire. And they made the mistake on the end of the second day of dynamiting a huge chemical warehouse . . . and that was just pyrotechnics plus."

By the second day, a huge flood of refugees was fleeing the fire, forming tent camps outside the city or riding ferryboats across the bay. A woman named Rosa Barreda, who lived with her mother, wrote to a friend:

> Many burned-out people passed our house with bundles and ropes around their necks, dragging heavy trunks. From the moment they heard that fatal,

> heart-rending sound of the trumpet announcing their house would be burned or dynamited, they had to move on or be shot. As the sun set, the black cloud we watched all day became glaringly red, and indeed it was not the reflection of our far-famed Golden Gate sunset.

Even neighborhoods that had not been severely damaged by the quake burned to the ground. After 4 days, 80 percent of San Francisco had been destroyed, and half of its population of 400,000 was homeless. The official death tolls were variously given as 300 to 700, but they never counted the Chinese or Latinos living in the slums, so the real count is thought to be about 3,000.

Almost as soon as the quake and fire were over, the city boosters and major businessmen vowed to rebuild, bigger and better than before. There were ambitious plans to redo the entire street grid and make it a more modern city, but in the end it was rebuilt on the old plan, with some wider streets and a lot more buildings with fireproof construction. They still had not learned the lesson of the need for steel reinforcing rods in all their masonry, and even today there are many dangerous old buildings that have not been retrofitted.

Nine years after the quake, in an effort to show that it had completely rebuilt, San Francisco hosted the huge Panama-Pacific International Exhibition, a world's fair in honor of the completion of the Panama Canal. To create the fairgrounds, the builders dumped much of the earthquake debris into the bay as landfill. After the fair, part of the fairgrounds became Golden Gate Park, but some of the new land was used to build the Marina District. This was the area that shook the most and suffered the most damage during the 1989 Loma Prieta quake down in Santa Cruz.

In addition to rebuilding, the city fathers also decided that mentioning the quake was bad publicity and would discourage investors, so they officially called it the "San Francisco fire of 1906." After all, fires were familiar and unavoidable events that had ravaged many major cities over the years, whereas earthquakes were scary and unpredictable. They did their best to suppress any mention of the quake in the media. They even fed false stories to the East Coast press, boasting that San Francisco had risen from its ashes in a mere week.

Of course, geologists saw this attempt at cover-up and public relations as both ridiculous and frustrating, especially when it censored legitimate scientific information the public should know. In 1908 John Branner wrote the following in the *Bulletin of the Seismological Association of America* (which was established as a result of the 1906 earthquake):

> A major obstacle to the proper study of earthquakes [was] the attitude of many persons, organizations, and commercial interests to the false position that the earthquakes are detrimental to the good repute of the West Coast, and that they are likely to keep away business and capital, and therefore the less said about them the better. This theory has led to the deliberate suppression of news about earthquakes, and even the simple mention of them.
>
> Shortly after the earthquake of April 1906 there was a general disposition that almost amounted to concerted action for the purpose of suppressing all mention of that catastrophe. When efforts were made by a few geologists to interest people and enterprises in the collection of information in regard to it, we were advised and even urged over and over again to gather no such information, and above all not to publish it. "Forget it," "the less said, the sooner mended," and "there hasn't been any earthquake" were the sentiments we heard on all sides.
>
> There is no doubt about the charitable feelings and intentions of those who take this view of the matter, and there is reasonable excuse for it in the popular but erroneous idea prevalent in other parts of the country that earthquakes are all terrible affairs; but to people interested in science, it is not necessary to say that such as attitude is not only false, but it is most unfortunate, inexcusable, untenable, and can only lead, sooner or later, to confusion and disaster.

THE BIRTH OF MODERN SEISMOLOGY

The 1906 San Francisco earthquake also represents another important landmark: it was one of the first quakes to be extensively studied by modern scientific methods, and the discoveries made in its aftermath formed the foundation of seismology. To undertake this study, the governor of California appointed a blue-ribbon panel of geologists and geophysicists. The State Earthquake Investigation Commission (SEIC) was set up under the leadership of Andrew Lawson, a legendary geologist and a faculty member of the University of California–Berkeley (see chapter 22). Lawson's commission looked at every aspect of the great earthquake and produced a pair of volumes, co-edited with the pioneering seismologist Harry Fielding Reid of Johns Hopkins University. Published in 1908, it was called *The Report of the State Earthquake Commission* and was more than 300 pages in length.

The authors of the volume included many of the best men in geology and geophysics of that time. One of them was Grove Karl Gilbert of the U.S. Geological Survey, whose many discoveries are still impressive a century later. He happened to be in Berkeley when the quake struck. Since he couldn't get into San Francisco for days, he traveled up and down the other parts of the Bay Area, mapping and photographing the fault displacement from up by Tomales Bay down to the southern Bay Area and San Juan Bautista (figure 23.2A). The most striking image is a fence line near Olema, California, which shows the huge amount of horizontal movement along the fault by the offset of the fence (figure 23.2B). Harry Reid used the 1906 earthquake to develop the elastic rebound theory of seismology, which is still accepted today. Other commission members went on to become the pioneers the geological community of California. J. C. Branner founded the Stanford Geology Department and was later appointed president of Stanford. H. O. Wood founded the Seismological Lab at Caltech. F. E. Matthes mapped the topography of many places, including Yosemite and other national parks. G. Davidson was the first president of the Seismological Society of America. The only foreign member, Fusakichi Omori, became one of the most celebrated seismologists in Japan.

This all-star lineup of geologists and seismologists compiled an impressive report, documenting not only the effects of the earthquake on structures, and how it was felt, but also detailing the physical effects, especially the fault offset and ground rupture. At that time, it was not yet proven that faults caused earthquakes, but the 1908 report settled that question once and for all. The Commission traced the rupture of the fault all the way from Point Delgada north of Tomales Bay, down to San Juan Bautista, and even mentioned effects as far south as Whitewater Canyon near Palm Springs. Although they didn't connect all these activities to the same San Andreas fault, or recognize that most of its motion was strike-slip (a fault that moves only in the horizontal plane), rather than vertical offset, they did recognize the California Coast Ranges were permeated by faults, including the Hayward fault beneath Oakland and the San Jacinto fault west of Palm Springs. They also documented many related phenomena, such as coseismic landslides and uplifted and downdropped crustal rocks; made maps of damage to structures; and wrote descriptions and analyses of the few seismographs then available and accounts of previous California earthquakes. And part 2 of the report, authored mainly by Reid, proposed not only the elastic rebound

Figure 23.2 ▲

Photographs of damage in the northern San Francisco Bay Area taken by Grove Karl Gilbert after the 1906 quake: (*A*) cracks in the ground (Gilbert's wife for scale); (*B*) the famous offset fence near Olema, California. (Courtesy of U.S. Geological Survey) theory of earthquakes but laid the foundation for the geophysics of earthquakes. In short, the SEIC report was the foundation for much of modern seismology.

Since the 1908 study, a lot more has been learned as seismological techniques have improved. The most widely accepted estimate for the size of the earthquake is a moment magnitude of 7.8. The main shock epicenter occurred about 3 kilometers (2 miles) offshore, near Mussel Rock. It ruptured along the San Andreas Fault both northward and southward for a total length of 477 kilometers (296 miles).

In 2006 the geological community commemorated the centennial of the San Francisco earthquake with many scientific meetings and publications. The overwhelming message from the assembled scientists was that the danger was not past. The 1989 Loma Prieta earthquake occurred on a stretch of the San Andreas fault just south of the 1906 break and caused severe damage in San Francisco (including breaking spans of the Bay Bridge and causing many houses in the Marina District to sink into the ground and burn). The major faults along the East Bay, especially the Hayward fault, are considered even more dangerous. In short, the earthquake hazards of San Francisco are not over. A quake just as big is long overdue, and it would be much more catastrophic than it was in 1906, due to the huge increase in population and infrastructure since then.

EARTHQUAKE MYTHS

When you mention the word "earthquake" and "fault" to anyone (especially in California), the first one that comes to mind is the San Andreas fault. Thanks to years of publicity from its great earthquakes, plus terrible movies full of bad science, it is the best-known fault in the world.

Unfortunately, what most people think they know about California earthquakes and the San Andreas fault is simply not true. First of all, California is *not* going to fall into the sea. The fault is moving horizontally, with the western side (Pacific plate) moving northwest relative to the eastern side (North American plate). Each time the fault moves, the Pacific plate slips northwesterly, on average about as fast as your fingernails grow. This means that it sometimes moves just a few inches, but other times it jumps over 10 meters (33 feet). In 50 million years, Los Angeles will be next to San Francisco, and eventually the entire block will plow into southern Alaska.

Second, the fault is nothing like it is portrayed in sensationalistic, unscientific movies like the 2015 movie *San Andreas* with Dwayne Johnson or the 1974 movie *Earthquake* with Charlton Heston. There will be no big tsunamis (the San Andreas and related faults are almost always on land, so they won't displace seawater to make a tsunami), no shaking that lasts for more than a minute, and none of the other spectacular but fictional effects seen in movies like these. There will be no great chasm in the ground with lava deep in the crack, like in the 1978 *Superman* movie with Christopher Reeve. The fault line is just marked by a long straight valley, and it's not apparent where it is until after it breaks. The only chasms seen after quakes are those caused by landslides opening up gaps between blocks, usually far from the original fault line. Faults like the San Andreas slowly grind up the rocks on each side, making them erode more quickly, so a straight valley (often with a few small sag ponds where the fault line traps water) is all you can see. Geologists in California see this pattern so often that any straight linear valley is assumed to be a fault unless proven otherwise.

The list of myths and misconceptions goes on and on. There is no such thing as "earthquake weather." A careful analysis of quakes in almost any regions shows they can happen at any time and any season, with no preference for hot weather, cold weather, or any particular time of day. This is because earthquakes occur many miles underground, while the daily fluctuations of temperature are not felt more than a few feet below the surface.

Even more problematic are the overreactions and irrational fear of earthquakes, especially in the United States. Most people fear quakes more than any natural disaster, yet you are more likely to die of a lightning strike or a snakebite than be killed in a quake. In the United States, earthquakes kill fewer people (fewer than 6 per year) than those rare events like lightning or snakebites. By far the most deadly disasters are the commonplace ones, such as heat waves and severe blizzards, which are by far the worst killers in the United States. Closely following these catastrophes are hurricanes, tornadoes, and floods. Yet people will go out into blizzards and do stupid things that kill them during heat waves. They are not paralyzed with fear by these events, yet act irrationally when it comes to a far less deadly disaster such as a quake.

Why is this? One reason is the fact that unlike other events (which are weather related), earthquakes are totally unpredictable. There are quacks who claim to be able to predict quakes, but over 50 years of experience has taught seismologists that no two quakes are alike, and precursors that might warn us of one type of quake are useless for another with no precursors. By contrast, we can watch the weather forecasts for blizzards, hurricanes, tornadoes, and floods and make some preparations, so these phenomena are less frightening. The other factor is the deep psychological shock of finding out that the terra firma beneath our feet is not in fact so firm, which shakes us to our core.

This is not to say that there are no places where earthquakes are deadly. In the less developed or older parts of the world, most of the construction is antique building made of simple unreinforced brick and mortar, the worst possible construction in an earthquake zone. Both for lack of knowing better and lack of other options, such as wood framing (the best possible structures) or steel framing, people in much of the quake-prone world just keep rebuilding these death traps. This is why you hear of so much loss of life when there are big quakes in Turkey, Iran, Armenia, Nepal, China, and Italy. But in California, the Field Act (passed right after the 1933 Long Beach quake) makes such construction illegal, and the only brick-and-mortar structures in the state must be reinforced by steel rods running through them, so they hold together during the shaking. That's why you should not lose sleep over dying in an earthquake in California or just about anywhere in the United States. Instead, you should worry about that heat wave or blizzard, or just driving in traffic! Those events are far more likely to kill you.

THE SAN ANDREAS FAULT

Almost none of this information was known back in 1906 when the San Francisco quake struck and seismology was in its infancy. It's a measure of how much the science has grown, and how much has been learned, that we now have a tremendous amount of information about faults and earthquakes amassed in the 111 years since the quake.

The San Andreas fault was first recognized in 1895 by Andrew Lawson, pioneering geologist at University of California Berkeley. He is the same person who headed the commission in 1906 that studied the San Francisco quake. He named it not after Lake San Andreas (as is commonly claimed) but after the San Andreas Valley in which the lake sits (then called Laguna de San Andreas). After the 1906 Lawson Committee documented the movement during the San Francisco quake, the fault was recognized not only in the Bay Area, but traced as far as Southern California, which had shaken to a lesser degree.

Over the next few years, geologists mapped more and more segments of the San Andreas fault, but considered it no different from any other fault, except that it was much longer (figure 23.3A). Eventually, it was shown to extend more than 1,300 kilometers (800 miles) from Cape Mendocino and the Gualala area, then offshore until it reaches Point Reyes, then diagonally across the southern part of San Francisco (mostly through Daly City), before trending just west of Silicon Valley, under San Juan Bautista, and then down the Coast Ranges from Hollister to Parkfield to the Carrizo Plain (figure 23.3B). From there it curves almost east-west across the northern flank of the Transverse Ranges, then down between the San Gabriel and San Bernardino Mountains at Cajon Pass, before continuing through San Bernardino, Banning, Palm Springs, and finally down into the Salton Trough and the Mexican border.

During this early mapping, the biggest revelation was that the San Andreas Fault was responsible for not only the 1906 San Francisco quake, but other important events. The biggest such episode was the 1857 Fort Tejon earthquake, which broke a huge segment of the fault from central California all the way down to Fort Tejon (where the Grapevine and Interstate 5 cross the fault at Tejon Pass), about 350 kilometers (220 miles) in total. The Pacific plate jumped 10 meters (33 feet) north in just seconds, completely rupturing the roads through the canyons and passes—and many more highways and other structures cross this segment of the fault today. It has been estimated to have been a magnitude 7.9 earthquake, roughly the same size as the magnitude 7.8 event in San Francisco in 1906—but it was much less deadly, because it did not hit any large cities. In some cases, the shaking may have lasted almost 3 minutes, much longer than in San Francisco.

In 1857 most of southern California was sparsely settled, with just a few Army soldiers in the cavalry fort at Fort Tejon (where all the adobe buildings

Figure 23.3 ▶

The San Andreas fault in California: (*A*) A map showing major locations along the fault line. (*B*) From the air, the fault is easy to spot on the Carrizo Plain, where it forms a long straight scar along the ground with hills crumpled on each side. ([*A*] Redrawn from several sources; [*B*] Courtesy of Wikimedia Commons)

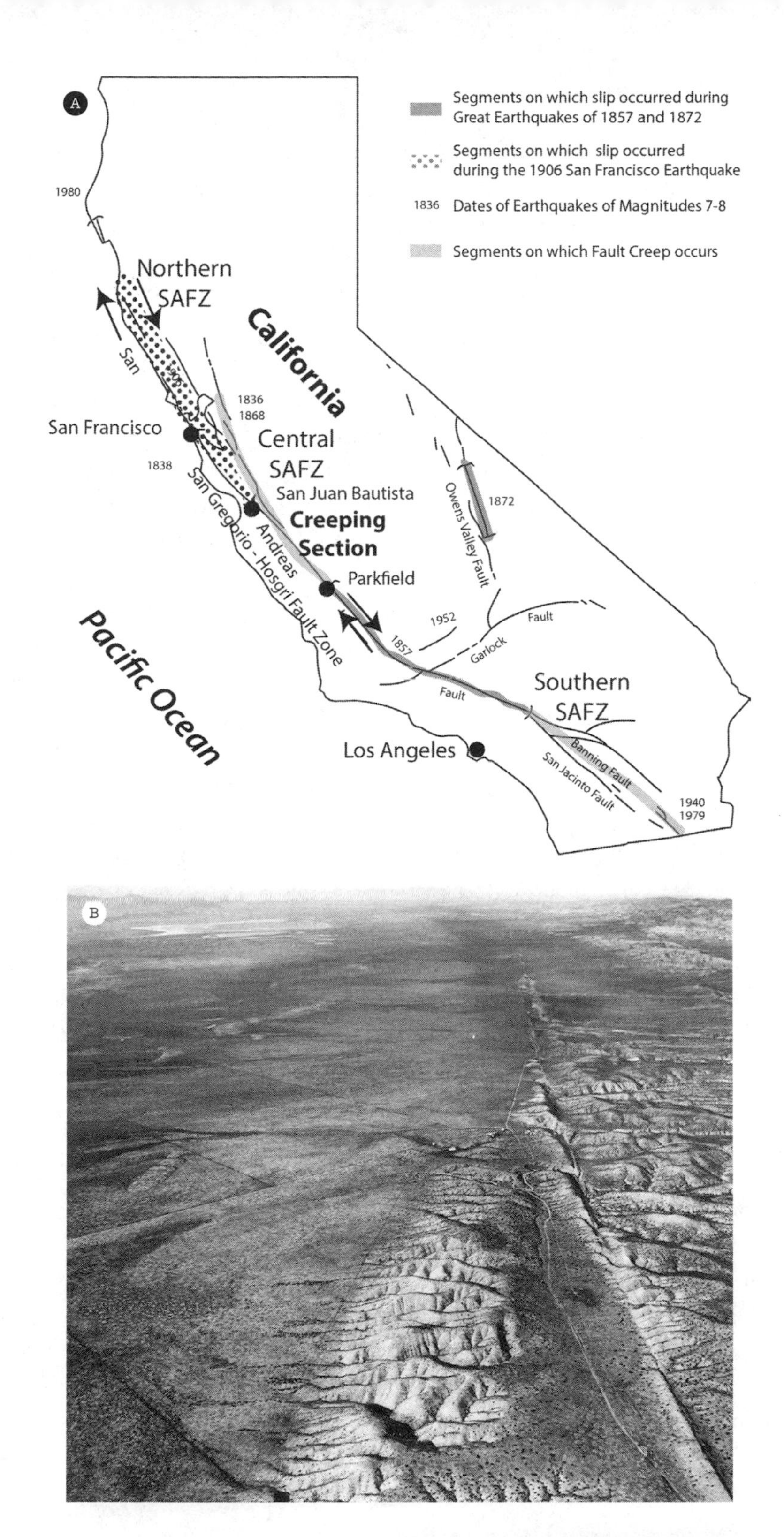
A
Segments on which slip occurred during Great Earthquakes of 1857 and 1872
Segments on which slip occurred during the 1906 San Francisco Earthquake
1836 Dates of Earthquakes of Magnitudes 7-8
Segments on which Fault Creep occurs
1980
Northern SAFZ
California
San
1906
1836
1868
San Francisco
Central SAFZ
1838
San Juan Bautista
Creeping Section
San Gregorio - Hosgri Fault Zone
Andreas
Owens Valley Fault
1872
Parkfield
Pacific Ocean
1952
Fault
1857
Garlock
Fault
Southern SAFZ
Los Angeles
Banning Fault
San Jacinto Fault
1940
1979
B

collapsed), and small populations in a few of the old Spanish and Mexican towns along the coast. Nevertheless, many masonry buildings collapsed, including some of the historic Franciscan missions all the way from Santa Cruz to Ventura to San Gabriel Mission just east of modern downtown Los Angeles. Only two deaths were reported, but this was mostly due to tiny populations living in small buildings at the time.

Since then, this stretch of the San Andreas fault has been quiet—too quiet (as they say in the movies). Most seismologists regard this fault segment to be "locked" so it cannot creep along with many small earthquakes (as happens in the Central California Coast Ranges and down in the Salton Trough), but instead builds up stress until it releases with a huge quake all at once. In the 1970s and 1980s, seismologists dug trenches across old pond deposits that were deposited across the fault valley for centuries. They were able to find layers going back almost 2,000 years at Pallett Creek. They could see layers in the trench cuts that were broken by older quakes, then covered by unbroken layers after the quake. By radiocarbon dating of pieces of charcoal in the layers, they could bracket the age of the quakes that occurred between the layers and get a history of quake activity on the fault. Based on these analyses, seismologists showed that that the deposits covered the last 2,000 years of activity on this stretch of the San Andreas. The first round of dating showed a recurrence interval of about 137 ± 8 years; a later study suggested a recurrent interval of 145 ± 8 years. Consider that in 2017 it will have been 160 years since the 1857 Fort Tejon quake, and you can see why seismologists are very worried about this stretch of the San Andreas. When it does slip, it will probably be offset about 10 meters (33 feet) in seconds, just as in 1857. It will be the "big one" that all Southern Californians have been awaiting and fearing for so long.

STAGGERING SLIP

In the years since the 1906 San Francisco quake, geologists saw the San Andreas fault as merely a long structure with at most a few tens of meters of offset over the long term. It was a very long fault, but its slip didn't seem out of the ordinary. But this simple idea was dealt a great blow by the most fundamental of geologic skills: plain, old-fashioned field mapping. And the men who made this discovery were among the most amazing geologists of all time.

One of these men was Tom Dibblee who spent more than 60 years doing geological field mapping. In 2004, he died at the age of 93, and was still mapping in his late 80s, although not moving as fast as he did when he was a young man. I was fortunate to meet him and chat with him several times in my career. A descendant of the original Mexican comandante of the presidio of Santa Barbara, Tom grew up on Rancho San Julian in the hills of western Santa Barbara County. When an oil geologist came to his family ranch to find possible oil-bearing structures, he got young Tom (then a high-school student) interested in geology. Tom got his degree in geology from Stanford in 1936, then worked for the California Division of Mines and Geology, publishing many reports on mercury deposits and other projects. Then he worked for Union Oil Company and Richfield Oil (now ARCO), and found large oil fields through his pioneering field mapping of the Temblor, Caliente, San Emigdio, and Southern Diablo Ranges; Carrizo Plain; the Cuyama, Salinas, and Imperial Valleys; the Santa Cruz Mountains; the Eel River area; and areas in western Oregon and Washington. Tom joined the U.S. Geological Survey in 1952 and was assigned to basic mapping in the Mojave Desert until 1967, where he covered more outcrops and made more discoveries than anyone before him. Once that project was done, he kept on mapping up and down the Coast Ranges of California, so after his "retirement" from the survey in 1977, he continued to map about 7,800 square kilometers (3,000 square miles) of coastal California for the U.S. Forest Service. Altogether, this one man mapped almost one-fourth of California, or about 100,000 square kilometers (40,000 square miles), more land than any human has ever mapped—or ever will map.

There are many legends about Dibblee and his techniques. His stamina was phenomenal, and even in his 70s and 80s, he could walk faster and farther than men one-third his age. He was only interested in reconnaissance mapping, so he did not walk out and see every square foot of a quadrangle, but mapped from roadcuts and the tops of mountains and ridges, where he could see the big picture. Later geologists often found that the details were more complex that Tom had realized, but this was a necessary compromise—he was focused on the big picture, not the details.

Most of the time, he was mapping in extremely remote areas, camping out with enough food and water to last him a week. Each night his campsite was his beat-up old car. He slept sheltered from the wind on the car seat with one door open and a board extending outward on which to rest his legs.

This simple camping arrangement, right in the heart of the field area (no driving back many miles into town for a hotel), enabled him to cover a lot of ground at little expense. He was notoriously frugal, even for someone who had to work on a shoestring budget. One of his Richfield Oil supervisors, stunned that he submitted an expense account for only $14.92 for an entire mapping project, said that he couldn't imagine feeding himself on such a small amount. Tom replied, "Oh, I find lots of things I like to eat up in the hills." Whatever the merits of his preliminary mapping, every geologist in California works in his shadows and usually starts with a copy of Tom's maps to figure out what is known and what needs to be solved. Fortunately, after his death his many geologist colleagues and supporters set up the Dibblee Geological Foundation, which keeps all his colored maps in print, so they can be ordered online at any time.

Another legendary California geologist was Mason L. Hill, known as "Mase" to his friends. Growing up in Pomona, California, he attended Pomona College, where legendary geologist A. O. "Woody" Woodford hooked him on geology as a profession. After peeling potatoes as Woody's field assistant, he graduated in 1926, then worked for the Black Hawk Gold Mine, then for Shell Oil, before getting graduate degrees from Claremont Graduate School and the University of California–Berkeley. There he wrote the first complete study of the geology of the San Gabriel Mountains. He then moved on to University of Wisconsin, where he got his doctorate in 1934, specializing in the mechanics of faulting.

In 1936 Hill started working for Richfield Oil, where he met Tom Dibblee, and the two collaborated many times. Hill would use Tom's maps to decipher the "sense of slip" on poorly exposed faults—and in the process, found many oil fields beneath invisible thrust faults. He continued to work for Richfield the rest of his career, reaching the post of chief geologist before he retired. He coined the standard terminology of strike-slip faults, made some of the first discoveries in Alaska's North Slope, and was a major contributor to the *Geology of Southern California*, Bulletin 170, published by the California Division of Mines and Geology in 1954. He finally retired in 1969, after many years of service not only to Atlantic Richfield (now ARCO), but also to many professional organizations.

Between them, Dibblee and Hill had seen more California geology than any human ever did or probably ever will. By 1953 they had reached an extraordinary conclusion, which they published in a landmark paper:

the San Andreas fault had moved hundreds of miles since the Jurassic (only 140 million years ago). Most geologists before them argued that its offset was only a few miles at best.

How did Hill and Dibblee arrive at this startling conclusion? During his mapping, Dibblee had discovered that there were rock units on one side of the San Andreas fault that clearly matched with others on the opposite side—but they were many miles apart (figure 23.4). Hill, in turn, used his skill in fault analysis to recognize the signs of how faults like this moved. For example, they found matches of rocks that were 100 kilometers (65 miles) apart since the late Miocene (only 7 Ma), 280 kilometers (175 miles) apart since the early Miocene (only about 20 Ma), and roughly 480 kilometers (300 miles) since the Late Jurassic (about 150 Ma). If you could restore the block of California west of the San Andreas fault to its

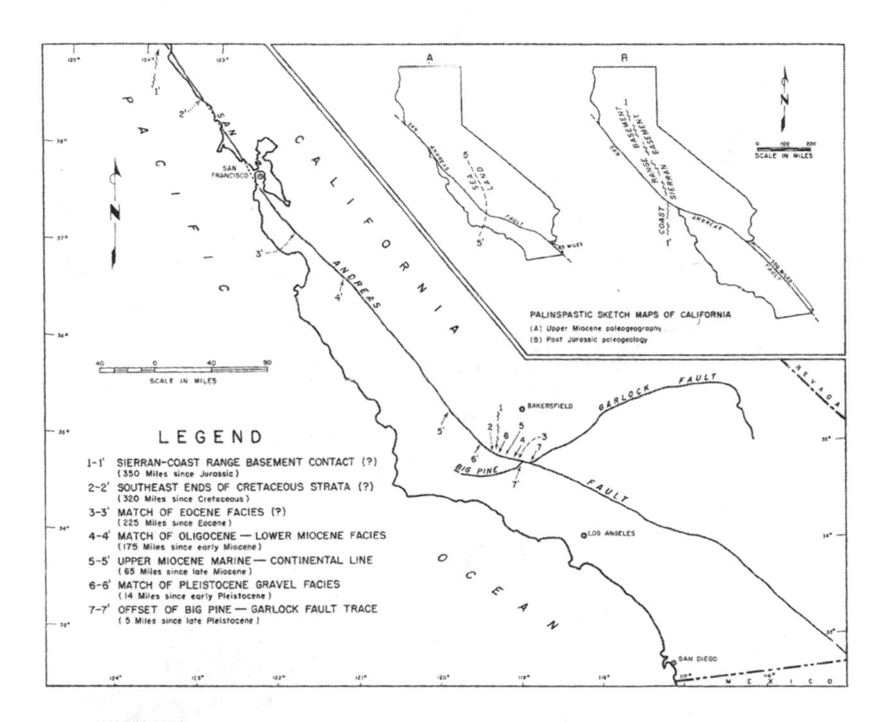

Figure 23.4 ▲

The displacement of various matching points along the San Andreas at different times in the geologic past, showing that in the Jurassic (*top inset*), most of western California was south of the current Mexican border. (Courtesy of the California Division of Mines, *Geology of Southern California*, Bulletin 170 [1954])

Jurassic position, you would see it had started well below the modern Mexican border and has moved enormous distances.

Naturally, such an outrageous idea was immediately challenged by many other geologists. They doubted that the similarities between the offset units on each side were that clear-cut or that the dating of these rock units was good enough. But in the years since 1953, more and more matches were found that confirmed Hill and Dibblee's daring suggestion. For example, the spectacular rocks at Pinnacles National Park in the central Coast Ranges (figure 23.5) are volcanic lavas that erupted around 23 Ma. They

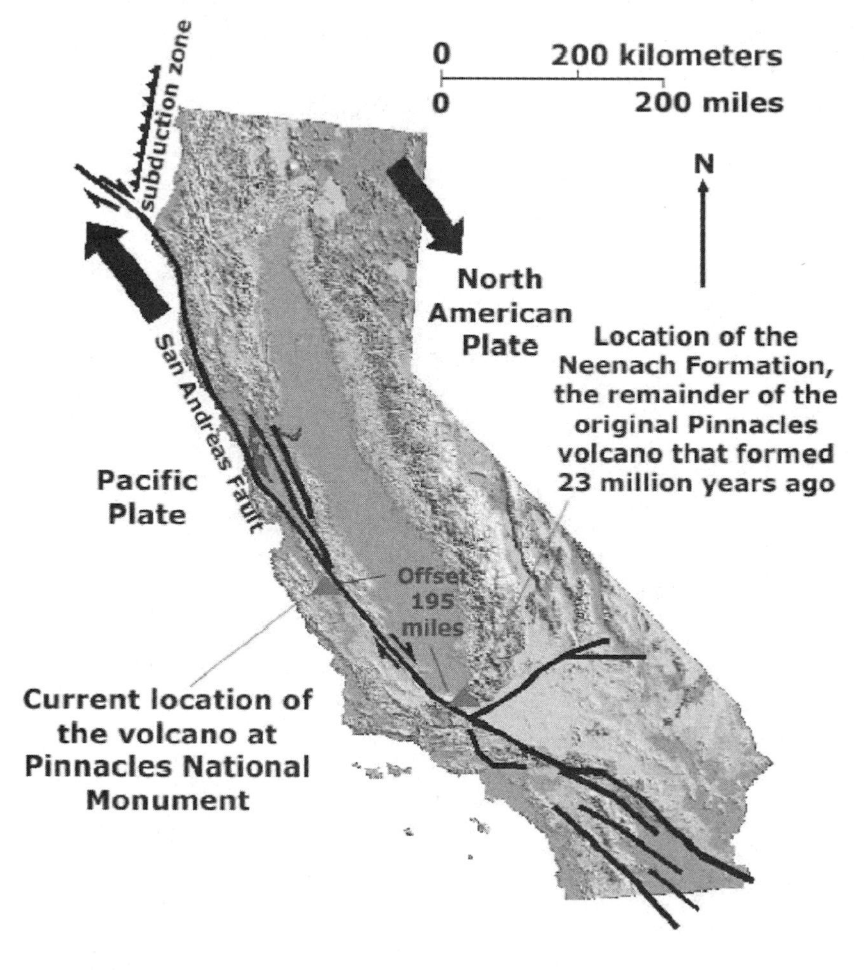

Figure 23.5 ▲

Displacement of the rocks at Pinnacles National Park, with their match down in the Mojave Desert in the Neenach volcanics. (Courtesy of U.S. Geological Survey)

are chopped off on their northeast side by the San Andreas fault, and their match is the Neenach volcanics in the western Mojave Desert west of Palmdale, over 313 kilometers (195 miles) away. They moved this distance in only 23 million years.

The Eocene Butano Sandstone north of Santa Cruz is matched by the Point of Rocks Sandstone in the Temblor Range, giving a distance of about 354 kilometers (220 miles) in about 40 million years (figure 23.4). Almost the same offset occurs between the Eocene Pelona Schist in the northern San Gabriel Mountains and its counterpart, the Orocopia Schist, east of the Salton Sea. The Paleocene rocks of the La Panza Range in central California match the San Francisquito Formation in the Transverse Ranges (figure 23.4).

As you get to even older rocks, the offset becomes even more amazing. The Cretaceous rocks of the Gualala block up by Point Arena match rocks in the Transverse Ranges, proving that they have moved 514 kilometers (320 miles) in about 100 million years. And the Jurassic basement beneath the Gualala block matches the Sierran basement, suggesting a transport of 560 kilometers (350 miles) in about 140 million years.

This amazing story of how far the western half of California was sliding seemed very hard to swallow in the 1950s and 1960s, even though the evidence grew stronger and stronger the more geologists looked. No one could imagine such huge horizontal motion on faults, or blocks sliding hundreds of miles or kilometers on a fixed continental crust that does not move. But everything changed with the birth of plate tectonics.

TRANSFORMED INTO A TRANSFORM

In the early days of plate tectonics, there was a cross-fertilization of many creative minds, mostly at a handful of cutting-edge institutions, such as Cambridge University, Lamont-Doherty Geological Observatory, Princeton University, and Scripps Institution of Oceanography. In 1965, all of them happened to be at Cambridge. Canadian geophysicist J. Tuzo Wilson took a sabbatical year at Cambridge at the same time that Frederick Vine and Drummond Matthews were solving the problem of seafloor spreading and magnetic anomalies (chapter 21), Harry Hess was visiting Cambridge from Princeton, Sir Edward Bullard was improving the fit of the continents and figuring out the physics of how crustal plates moved on a spherical globe, and brilliant minds like Dan McKenzie, John Sclater, and Robert Parker were working on the key problems of plate tectonics as Cambridge research students. Wilson had already converted to the idea of moving continents as early as 1960, and he published a landmark paper in 1963 showing that the Hawaiian Islands were formed as the Pacific plate slipped over a fixed hot spot in the mantle. The current Big Island is over the hot spot now and still erupting at Kilauea Volcano, but each of the islands to the northwest is progressively older, representing earlier eruptions that stopped when they slid past the hot spot. In chapter 17, we discussed how Wilson recognized that the displaced trilobites and other fossils on each side of the Atlantic were evidence of a proto-Atlantic that had closed, then re-opened along a different line.

But in 1965 Wilson was part of the first theoretical developments of how plates moved with respect to one another. Inspired by ideas from Bullard, McKenzie, and also Australian geologist S. Warren Carey (who advocated an expanding earth), Wilson recognized that there were three basic types of possible plate boundaries. Vine, Matthews, and Hess had all developed the idea of seafloor spreading, places where plates pull apart (chapter 21). This normally occurred at mid-ocean ridges and was becoming better understood by 1965. Holmes, Hess, and several others (chapter 22) had also postulated the idea of crustal plates coming together and one plunging beneath the other.

But Wilson realized that if new plate was erupting out of a mid-ocean ridge on one part of the globe and slowly sliding to its destruction in a trench on another part of the globe, there must be a third kind of plate boundary where the plates were neither separating nor colliding, but just sliding past one another. These were named transform faults by Wilson, because they transport or "transform" the plate from one place to another. Their motion is mostly horizontal, strike-slip motion, with almost no vertical movement, and very little compression or separation. Looking at the map of plate boundaries that was quickly emerging, Wilson saw that most of the mid-ocean ridges had short fault segments that offset the ridges. These were the first examples of transform faults he discussed. He also showed why mid-ocean ridges must have these transform faults to accommodate the difference in movements of the plates as they rotate on the spherical surface of the earth.

But he looked even harder and noticed that there were many examples of huge strike-slip faults around the globe that had been a mystery. In every case, these transforms were the link connecting two other types of plate boundaries, such as offsetting two spreading ridges, or connecting a ridge to a trench. He worked out all the possible geometries of three plate boundaries interacting and gave real-world examples for many of them.

Wilson then looked at the mystery of mysteries, the San Andreas fault (figure 23.3A). Sure enough, he could show that it began at the north end of the East Pacific Rise, a spreading ridge that ran up from the coast of South America and then down the center of the Gulf of California. The San Andreas took up the spreading of crust from that ridge, linking it to another ridge and transform system up on Cape Mendocino.

The outrageous slip movements suggested by Hill and Dibblee in 1953 suddenly made sense. They were the direct result of plates sliding across the surface of the earth, and the San Andreas was the fault that takes up the slip of the Pacific Plate on its way northwest to subduct beneath the Aleutian arc and all the subduction zones of the western Pacific.

Mystery solved.

FOR FURTHER READING

Collier, Michael. *A Land in Motion: California's San Andreas Fault*. Berkeley: University of California Press, 1999.

Cox, Allan, ed. *Plate Tectonics and Geomagnetic Reversals*. San Francisco: Freeman, 1973.

Cox, Allan, and R. B. Hart. *Plate Tectonics: How It Works*. New York: Wiley-Blackwell, 1986.

Dvorak, John. *Earthquake Storms: The Fascinating History and Volatile Future of the San Andreas Fault*. New York: Pegasus, 2014.

Hough, Susan E. *Finding Fault in California: An Earthquake Tourist's Guide*. Missoula, MT: Mountain Press, 2004.

Molnar, Peter. *Plate Tectonics: A Very Short Introduction*. New York: Oxford University Press, 2015.

Oreskes, Naomi. *Plate Tectonics: An Insider's History of the Modern Theory of the Earth*. New York: Westview, 2003.

Winchester, Simon. *A Crack in the Edge of the World: America and the Great California Earthquake of 1906*. New York: Harper Perennial, 2006.

Yeats, Robert S., Kerry E. Sieh, and Clarence R. Allen. *Geology of Earthquakes*. Oxford: Oxford University Press, 1997.

24 THE MEDITERRANEAN WAS A DESERT

MESSINIAN EVAPORITES

We are tied to the ocean. And when we go back to the sea, whether it is to sail or to watch—we are going back from whence we came.

—JOHN F. KENNEDY

FROM THE ASHES OF DISASTER . . .

By the late 1940s and early 1950s, the field of marine geology was growing by leaps and bounds. Oceanographic vessels from the Lamont-Doherty Geological Observatory in New York, Woods Hole Oceanographic Institution in Massachusetts, and Scripps Institution of Oceanography were in the early phases of more than 30 years of voyages around the world, collecting every bit of information about the oceans that they could. Year after year, the ocean began to yield its secrets. Its temperature and chemistry were routinely measured by water samples from every depth. Its depth and shape were gradually deciphered by echo sounding and sonar. Its shallow structure beneath the surface was revealed by throwing dynamite off the fantail of the boat and measuring the way that sound waves bounced back from the layers beneath the seafloor. Long steel tubes called piston corers were plunged into the sediments on the sea bottom, yielding cylindrical cores measuring up to 10 meters (33 feet) in length—but no longer. Whenever they tried to drop much longer sections of steel coring pipe, they ran into problems, so their sedimentary record was relatively short and recent, and mostly represented the last million years of Ice Age deposits.

Meanwhile, for over 60 years seismologists had been deciphering the structure of the earth's interior from the ways that seismic waves from big earthquakes showed up on their seismographs. Considering the way in which the seismic waves were bent, they had determined the structure of the core and mantle, the temperature and density of each of the layers, and that the outer core was fluid.

One of the first discoveries was made in 1909 by a Croatian seismologist, Andrija Mohorovičić. Looking at seismograms that showed waves coming up from the upper part of the earth's mantle and crust, he noticed that there was a dramatic change in speed between waves that traveled only in the crust and those that traveled down into the mantle and then back through the crust before being recorded by seismographs. Calculating the depths from which these seismic waves emanated, Mohorovičić determined that there must be a sharp difference in density, and therefore a distinct boundary, between the crust and the uppermost mantle. Later seismologists confirmed this discovery, and this sharp break between crust and mantle came to be known as the Mohorovičić discontinuity. The name is such a mouthful that geologists call it the "Moho" for short.

By the late 1940s, seismologists had determined the depth of the crust in many parts of the world. They found that oceanic crust was relatively thin, with only about 10 kilometers (6 miles) of crust overlying the mantle. Continental crust, on the other hand, was at least 5 to 15 times thicker, with thicknesses ranging from 50 to 150 kilometers (30 to 100 miles). Thus, if we ever wanted to drill down to get a sample of the mantle, the best place would be to drill through the thinnest part of the oceanic crust.

This was an idea that interested the famous marine geophysicist Walter Munk (figure 24.1). Born in Vienna in 1917, Munk was sent to school in New York in preparation for a career in banking. Young Walter disliked banking and enrolled at Columbia University. He transferred to Caltech and earned a bachelor's degree in physics in 1939 and a master's degree in geophysics in 1940. He started doing graduate research at Scripps, but when World War II came along, he volunteered. The U.S. military, however, saw his oceanic expertise as a valuable asset, so Munk served in a research group that helped determine the ideal tidal and surf conditions for the Allied landings in North Africa and the Pacific and for the D-Day invasion in Normandy. In 2009 Munk commented, "The Normandy landing is famous because weather conditions were very poor and you may not realize it was postponed

Figure 24.1 ▲

Walter Munk receiving the Crafoord Prize, the "Nobel Prize" of geology, in 2010. (Courtesy of Wikimedia Commons)

by General Eisenhower for 24 hours because of the prevailing wave conditions. And then he did decide, in spite of the fact that conditions were not favorable, it would be better to go in than lose the surprise element, which would have been lost if they waited for the next tidal cycle coming in two weeks." He was also important in advising the military on the prevailing tidal currents and wind conditions that would affect the H-bomb test on Bikini Atoll.

After the war, Munk returned to graduate school, getting his doctorate at UCLA in 1947, and then joining the faculty at Scripps, where he spent the rest of his career. He was instrumental in founding the Institute of Geophysics and Planetary Physics at Scripps and took part in many oceanographic cruises. Of his many discoveries, one of the most important was deciphering how the trade winds of the earth drive huge spirals of water circulation in the temperate and tropical oceans, known as gyres. He also demonstrated that the moon was tidally locked with the earth, so that one side always faces the earth. His best-known accomplishment, however, was developing the science of surf forecasting.

His most ambitious project (in collaboration with Harry Hess of Princeton) was an attempt to reach the mantle by drilling through the ocean floor. In 1952 Hess and Munk established a group of scientists (including Maurice "Doc" Ewing, head of Lamont, and Roger Revelle, head of Scripps) called the American Miscellaneous Society (AMSOC), to help advise the government about the merits and feasibility of certain scientific projects. By 1956 these prominent men had recruited a consortium of oil companies (Continental, Union, Superior, and Shell, or CUSS) to develop an ambitious idea: drill through the oceanic crust from a floating vessel to reach the mantle. Eventually, the project was taken over by the still-young National Science Foundation (NSF), working in collaboration with the CUSS consortium. Dubbed "Project Mohole" (as in "hole to the Moho"), it was as ambitious a project as the space race, which was also launched in 1957 in response to the Soviets launching the first satellite, Sputnik.

After a practice run near Scripps, the first attempt was launched in March 1961 off Guadalupe Island, Mexico. They contracted with Global Marine of Los Angeles to use its pioneering ocean-drilling ship, *CUSS 1*. This ship was the first ever built by oil companies to drill the ocean floor in search of new oil deposits, something that no individual company could accomplish at that time.

The collaboration between the NSF and the CUSS consortium was mutually beneficial. Scientists reaped the rewards of the research results, while the oil companies exploited the newly available offshore resources. The drilling was challenging, with 3,600 meters (11,700 feet) of seawater to navigate before they reached the ocean floor. Five holes were drilled, the deepest of was 183 meters (601 feet). They sampled only 14 meters (44 feet) of oceanic crustal basalt at the very bottom of the hole. There was no intention to drill much deeper on that first voyage, as the entire exercise was intended as an experimental phase to see whether ocean drilling was possible. However, Project Mohole never got the chance to try again. Politics got in the way. When the responsibility for administering the project shifted to the NSF, AMSOC disbanded, and many difficulties arose, Congress decided to cut off funding. Mohole was viewed as a failure.

. . . GROW THE ROSES OF SUCCESS

Meanwhile, the difficulties of drilling down through hard basalt had revealed something else: the upper 159 meters (557 feet) of the first attempt at drilling had bored easily through soft oceanic muds and shells of plankton that were Miocene in age. These deep-sea cores gave the first long record back in time through the ocean sediments, something that ordinary coring by dropping steel tubes off the ship could not. Although Project Mohole had fallen out of favor and lost its funding by 1966, both the scientists and the oil companies came away from the experience with an understanding that drilling through layers of oceanic sediments would be easier and more valuable. After all, the oil companies were only interested in sedimentary rocks, since that is where the oil was formed and trapped; they had no interest in the hard basaltic lava beneath the sediment. The scientists realized that drilling sediments around the world could provide a detailed history of all the oceans and their changes through many millions of years. Unlike the sedimentary record on land, which is very incomplete and episodic, the muds and shells of plankton that rain down from the ocean surface to the seafloor provide a nearly continuous record of earth history over millions of years with few missing time intervals.

By June 1966, Scripps and a different consortium of oil companies formed a new project cosponsored by the NSF and private oil companies—the Deep Sea Drilling Project, or DSDP. By October 1967, they had begun building a newer, more advanced ship than *CUSS 1*, called the *Glomar Challenger* (figure 24.2). Its name came from the shipbuilder Global Marine, Inc. ("Glomar" in the trade). It also honors the famous HMS *Challenger*, a British sailing ship that traveled the world's oceans from 1872 to 1876 on the world's first true oceanographic expedition. Launched on March 23, 1968, the *Glomar Challenger* was 120 meters (400 feet) long, 20 meters (65 feet) wide, and could sail at 22 kilometers per hour (14 miles per hour) for up to 3 months. Topped by a 60-meter (200-foot) drilling derrick, it would drill in water depths of 6,100 meters (20,000 feet), and eventually it could send

Figure 24.2 ▲
The early oceanographic drilling ship *Glomar Challenger*. (Courtesy of Wikimedia Commons)

down a drill string that could drill through 800 meters (2,500 feet) of sediments on the sea floor.

The first two legs of the ship's history were shakedown cruises in the Gulf of Mexico to make sure everything worked. Leg 3 was the first true scientific project, and naturally they set out to test the hottest idea in geology in 1968: whether or not seafloor spreading was real. They sailed to the South Atlantic and drilled a series of cores on each side of the Mid-Atlantic Ridge. Sure enough, the sediments at the bottom of each core were older and older the farther they were from the ridge, proof that the seafloor was indeed spreading (see chapter 21).

By 1983 *Glomar Challenger* had been in almost continuous operation for 15 years, with 96 separate expeditions or "legs." It had logged 695,670 kilometers (375,632 miles) of sailing and drilled 624 holes in the seafloor, recovering 19,119 cores. In the process, it had obtained an amazing record of the history of the world's oceans that solved all sorts of mysteries, from the causes of the Ice Ages (see chapter 25) to the extinction that killed the dinosaurs (see chapter 20) to how the oceanic currents had changed and affected climate for the past 150 million years. Many people regard it as one of the most important projects in the history of science, and certainly the most important in marine geology and oceanography.

But the equipment on the *Glomar Challenger* was worn out and outdated, so the ship was retired and unceremoniously cut up for scrap. This was a shame, because as a scientific instrument, it was as historically important as the Mount Wilson telescopes used to explore the expanding universe or the cyclotrons used to develop all of modern nuclear physics. In 1985 the *Glomar Challenger* was replaced by a newer, more advanced ship, the *JOIDES Resolution*, which has since traveled over 572,574 kilometers (355,781 miles), completed 111 legs, and drilled 1,797 holes with over 35,772 cores recovered. Although the *JOIDES Resolution* is still sailing, it is now semi-retired after 32 years of service.

The current phase of the ocean drilling is carried out by a huge Japanese ship, the *Chikyu Maru* (Japanese for "Planet Earth ship") (figure 24.3).

Figure 24.3 ▲
The *Chikyu Maru* drilling ship. (Courtesy of Wikimedia Commons)

Its construction began in 2002, and it first sailed and commenced drilling operations in 2007. Unfortunately, it was damaged in the 2011 Japanese tsunami, when it broke loose from its moorings and collided with a pier. It is so enormous—with a length of 210 meters (690 feet), a width of 38 meters (125 feet), and a derrick that towers 130 meters (430 feet) above the ocean floor, taller than the Statue of Liberty or the Gateway Arch in St. Louis—that it has been nicknamed “Godzilla Maru.” It has a separate helicopter pad and a range of over 27,000 kilometers (17,000 miles). It carries more than 200 people, including a crew of 100. It can drill down farther than 10,000 meters. On September 6, 2012, the *Chikyu Maru* set a new world record by drilling down and obtaining rock samples from deeper than 2,111 meters (6,925 feet) below the seafloor near the Shimokita Peninsula of Japan in the northwest Pacific Ocean. On April 27, 2012, *Chikyu Maru* drilled to a depth of 7,740 meters (25,400 feet) below sea level, setting a new world record for deep-sea drilling. That record still stands.

In a final nod to Project Mohole, the *Chikyu Maru* is scheduled to drill down to the mantle sometime in the near future.

PUZZLE #1: SCYLLA AND CHARYBDIS

One of the most remarkable places in the world is the Strait of Messina, between Sicily and the Italian mainland. Less than 3.1 kilometers (1.9 miles) at its narrowest point, it has a major effect not only on marine life in the Mediterranean, but on terrestrial life as well. It is a narrow bottleneck with unusual marine life (including deep-sea viperfish occasionally brought to the surface) and an important flyway for nearly all the birds migrating from Africa through Sicily to Europe and back. During the spring migration, some 300 bird species are typically recorded, and a record 35,000 raptors were spotted during one migration.

The strait was famous in the ancient world. It is mentioned in the legend of Jason and the Argonauts and in Aesop’s writings. One of the oldest accounts of the strait comes from Homer’s *Odyssey*, which relates how Odysseus and his crew traversed the narrows. They faced the perils of a six-headed monster named Scylla on the Italian side and a gigantic whirlpool named Charybdis that sucked down ships on the Sicilian side (figure 24.4). The gap was so narrow that no ship could pass unmolested; the only choice was which danger to approach. On the advice of the enchantress Circe,

Figure 24.4 ▲

The mythology of steering between Scylla and Charybdis was once a common metaphor for making a decision between tough choices. For example, this British editorial cartoon from 1793 shows Britannia, the personification of Great Britain, sailing between Scylla and Charybdis, guided by Prime Minister William Pitt. It is subtitled, "The Vessel of the Constitution steered clear of the Rock of Democracy, and the Whirlpool of Arbitrary-Power." Pitt is steering the small boat, *The Constitution* toward a castle with a flag inscribed "Haven of Public Happiness." They are pursued by Richard Brinsley Sheridan, Charles James Fox, and Joseph Priestley, who are sharks, or the dogs of Scylla. (Courtesy of Wikimedia Commons)

Odysseus chose to sail near Scylla, losing a few sailors to the monster, but saving his ship from the whirlpool. In Homer's immortal words, Circe told Odysseus: "Hug Scylla's crag—sail on past her—top speed! Better by far to lose six men and keep your ship than lose your entire crew." Then, as Odysseus approaches the rock of Scylla, the monster's heads snake out and grab six men from the deck. As written in the *Odyssey*:

> *They writhed gasping as Scylla swung them up her cliff and there*
> *at her cavern's mouth she bolted them down raw—*
> *screaming out, flinging their arms toward me,*
> *lost in that mortal struggle.*

Thanks to the generations who have learned classical literature over the years, the phrase "between Scylla and Charybdis" has since entered the English language as an idiom for having to make a difficult choice between two bad options, similar to phrases such as "on the horns of a dilemma," "between a rock and a hard place," "between the devil and the deep blue sea," and "between death and doom."

Homer's myth was not entirely imaginary. There are many shoals of hidden rocks on the rugged stormy coast on the Italian side, which may have led to the legend of Scylla. A famous rock on the shore of Calabria is called the "Rock of Scilla," supposedly the perch and cave where the monster dwelled. Charybdis is no myth at all. Due to the odd circulation patterns and strong tidal flow through the narrow channel bottleneck, a large whirlpool frequently appears on the Sicilian side of the strait. It may not have been as big as the 23-meter (75-foot) Charybdis that was large enough to swallow a Greek trireme, but the whirlpool is real and still treacherous to boats today.

To geologists, "Messina" has a different connotation. In the 1800s, geologists first working in the area of Sicily around Messina found enormously thick deposits of gypsum and salt (figure 24.5). These are minerals that form by evaporation of water in lakes or in marine basins, so a few

Figure 24.5 ▲
Thick deposits of salt and gypsum interbedded with marine muds in the Strait of Messina. (Courtesy of Wikimedia Commons)

inches of salt or gypsum requires a lot of seawater to evaporate away to produce them. In some places, the salt and gypsum deposits are more than 1,500 m (5,000 feet) thick and represent huge volumes of seawater that must have evaporated. Sicilians had been mining these salt deposits since ancient times, when they were valuable commodities shipped all over the ancient world.

When geologists first studied these deposits, they could not explain how they formed. Nevertheless, in 1867 geologist Karl Mayer-Eymar did detailed research on them and found some important clues. In the layer just above the top layer of salt and gypsum, he found fossils that indicated the muds were formed in brackish lagoons, with water chemistry somewhere between freshwater and seawater. The fossils also established that these beds were latest Miocene in age. Immediately above the brackish layer was a deposit of deep-sea muds, formed by clear, cold waters of normal salinity, covering the deposits formed from evaporation at the surface. The thick evaporite sequence came to be the basis for the latest Miocene Messinian Stage in Europe, and the deep-sea muds above it for the earliest Pliocene Zanclean Stage.

Many other examples of extraordinarily thick upper Miocene salt and gypsum beds were subsequently found around the Mediterranean. But no one could explain how they had formed, and why there were so many evaporites formed at that time.

PUZZLE #2: THE GRAND CANYON OF THE NILE

The Nile River is a mighty force of nature. Arising from the highlands of Kenya and from Lake Victoria, it travels 6,853 kilometers (4,258 miles) to its mouth on the Mediterranean, the longest river in the world. Over the last half of its course, it travels through the harsh deserts of Sudan, Ethiopia, and Egypt, bringing life to regions without any water.

When huge rains occur at its headwaters, they send enormous floods down its valley, wiping out all the small villages on its floodplains. The floods also bring fresh silt and organic material to the floodplain, making them some of the richest soils in the world. Thanks to the Nile and its fertile floodplains, one of the world's oldest civilizations sprang up in ancient Egypt over 6,000 years ago. As the Greek historian Herodotus wrote, "Egypt is the gift of the Nile." For centuries Egyptians had to cope with the annual floods on the Nile in order to reap the benefits of its rich agricultural bounty. They did so with great success for thousands of years. The Egyptian calendar was also pegged to the Nile, with three distinct seasons: *Akhet* ("flood"), *Peret* ("growing season"), and *Shemu* ("dry season"). For thousands of years, Egypt was the "breadbasket" of the ancient world, producing enormous crops of grain, and in more recent years, the famed Egyptian cotton. Egyptians were resigned to the loss of life and property that those annual floods brought as well.

In 1954 the Egyptian military staged a coup against the corrupt monarchy of King Farouk and took over Egypt. At the head of the military was an ambitious officer, Gamel Abdul Nasser. An ambitious and charismatic leader, he soon thought of himself as the leader of all the Muslim countries in the Middle East. He also tried to maintain strict neutrality in his dealings with the United States, Britain, and their allies on one side, and the Soviet Union, China, and the Communist bloc during the height of the Cold War. At first he turned to the United States when he needed help, but President Eisenhower and CIA chief John Foster Dulles only wanted to give him weapons for defensive purposes, with U.S. military advisers attached. Nasser refused these conditions. When an Egyptian raid on Israel in 1955 further strained relations, Nasser turned to Premier Nikita Khrushchev and the Soviet Union for help.

In 1956, to raise money for a dam on the Nile, Nasser seized the Suez Canal for its revenues. This triggered a world crisis. The United Kingdom, France, and Israel invaded Egypt, and the world almost went to war over the Suez Canal, but in the end the United States and the Soviet Union brought pressure to bear and the forces were eventually withdrawn. Egypt has controlled the Suez Canal ever since.

Once things had cooled down in 1958, the Soviet Union promised to pay for the building of a Nile dam at Aswan, called the Aswan High Dam. The dam was begun in 1960 and completed in 1970. It had some significant negative effects. The most famous was the flooding of some amazing ancient Egyptian temples, such as Abu Simbel. By 1960 a rescue operation under UNESCO began to cut the enormous sculptures of Ramses II into blocks, moving them to higher ground (figure 24.6A and B) and then reconstructing the entire temple on the shores of the reservoir (now called Lake Nasser). The sites of Philae, Kalabsha, and Amada also had to be moved. Some of Egypt's relocated temples were given to the countries that helped

A

B

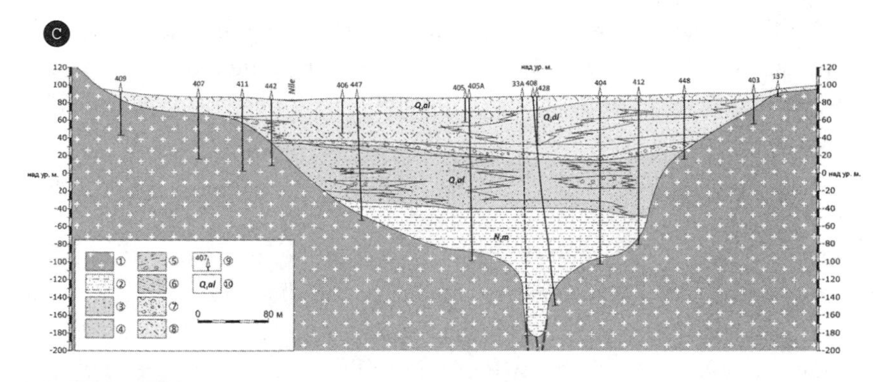

Figure 24.6 ◀ ▲

The effects of building the Aswan High Dam: (*A*) The temple of Abu Simbel, which had to be moved to higher ground. (*B*) Moving the pieces of the immense statues of Ramesses II. (*C*) The cross-section through the grand canyon beneath the Nile, showing the original Russian notation of the geologists who drafted the drawing. (Courtesy of Wikimedia Commons)

with the rescue effort, including the famous Temple of Dendur that now sits in the Metropolitan Museum of Art in New York.

The dam provided protection from floods and a steady source of water in case of droughts (which also plagued Egypt and the Sudan) as well as hydroelectric power, but at a price. The lack of annual flooding on the Nile meant that the soils were no longer replenished each year, and agriculture has suffered as a consequence. In addition, the lack of floodwaters and the high rates of evaporation in the Nile Valley have allowed the salts in the soil to percolate to the surface, because they are no longer diluted or washed away. Many important farming areas have become too salty to grow crops any more.

The lack of sediment has also meant that the Nile Delta is no longer building out, but instead is being eroded back by the actions of the Mediterranean Sea. Instead, the river sediment is trapped in the bottom of Lake Nasser, which is slowly becoming shallower and clogged up and eventually will be useless. The water of the Nile is no longer muddy, and the clear waters are now choked with algal growth, which affects drinking water quality all through Egypt. Fisheries in the Nile Delta and Mediterranean offshore region have collapsed. In short, many people think that the economic costs of Nasser's project outweigh any benefits it provides, especially as Egypt is now the most populous country in Africa and is straining under issues of poverty and political unrest.

But the construction of the Aswan High Dam had another surprising consequence. Soviet and Egyptian engineers picked the dam site because of the rugged narrows between sandstone cliffs on each side. Then they drilled boreholes across the valley to find the depth of the hard bedrock at the bottom so they could anchor the foundation of the dam. They drilled and drilled in the Nile Valley, but could not seem to find bottom, just more and more sediments filling the valley, deeper and deeper below the floodplain. By 1967 Soviet geologist Ivan S. Chumakov had discovered that the drill cores were full of Pliocene marine plankton from deep-water conditions. Finally, they hit bedrock, many thousands of feet below sea level! Drilling below Cairo showed that the bedrock was more than 2,500 meters (8,200 feet) below the surface of the floodplain. Further research, which entailed bouncing seismic waves off the subsurface layers, showed that there was a feature as big as the Grand Canyon beneath the surface. Without intending to, the dam engineers had discovered the Nile Valley is actually the fill of an ancient Grand Canyon (figure 24.6C), cut down 2,500 meters (8,200 feet) below present-day sea levels.

But what could cause the Nile River to cut down this far below modern-day sea levels? And why had it then completely filled with sediment after it was cut down so deep? Another mystery was discovered, and another isolated piece added to the puzzle.

PUZZLE #3: THE HOLE AT THE BOTTOM OF THE OCEAN

For decades, seismologists had been bouncing sound waves off the bottom of the Mediterranean Sea. They could see the layers of deposits of the basin as it sank down during the Miocene and Pliocene and earlier times. For most of the Mesozoic and Cenozoic, it was part of the great Tethys Seaway that ran from Gibraltar to Indonesia. Then, in the Miocene, it buckled down due to the pressure of being crushed between Africa and Europe as they collided (this collision was also responsible for raising the Alps). In 1961, however, the seismic profiles first revealed a distinct layer that gave a very "bright" reflection of intense sound waves. It became known as the "M reflector," and it was made of something that was very different in density than the sand and mud layers in the rest of the section, so early on they guessed it might be a salt deposit. It formed a blanket around the entire Mediterranean, hugging the seafloor but deeply buried beneath it, suggesting that it was once deposited across the entire sea bottom, then buried by younger deposits. In 1967 Italian seismologist Giorgio Ruggieri suggested that not only was the M reflector made of salt, but that it matched the salt beds long known from the Strait of Messina. He suggested that the entire Mediterranean had once dried up into a gigantic salt basin, and coined the term "Messinian salinity crisis." However, there was no direct evidence to confirm his hunch.

This, however, was clearly something worth investigating. Scientists proposed that the *Glomar Challenger*—only in the thirteenth leg of its surveying cruises and a mere 2 years after its first voyage—be assigned to drill the bottom of the Mediterranean and find out what the M reflector was—and whether Ruggieri was right. As in every DSDP voyage, Leg 13 carried a large scientific crew, headed by three scientists in charge. One of them was the lead micropaleontologist, Maria B. Cita. The second was Ken Hsü, a Chinese-born sedimentologist who had pioneered many areas of research, including figuring out what formed California mélanges (chapter 23). Now 88 years old, he has received almost every honor possible in geology for all his important discoveries, including the Wollaston Medal of the Geological Society of London and the Penrose Medal of the Geological Society of America.

The third was marine geophysicist William B. F. Ryan, whose focus was on the geophysics and depth profiling and coring. Bill Ryan was my marine geology professor in 1977 when I was a graduate student at Lamont, so I heard this story directly from him. He had been on dozens of Lamont oceanographic expeditions and pioneered much of the technology of echo sounding, oceanic surveying, deep-sea cameras and dredges, and other inventions that helped collect more data from the seafloor. A quiet, unassuming, buttoned-up kind of guy, he speaks slowly and softly, but his ideas and work ethic are unparalleled. Even though he speaks in a calm, quiet manner during class lectures, I still vividly remember him describing his experience mapping the abyssal plains of the world's oceans for the first time. They were so flat and so enormous, he told us, that the scientists would cruise for days over the abyssal plain with the echo sounder constantly giving the same depth reading.

Ken, Bill, and Maria all came to the project with different goals: to find the cause of the well-known drop in sea level across all of Europe in the late Miocene (Hsü); to find a good marine Miocene-Pliocene sequence and improve the biostratigraphic dating of marine rocks of the

Mediterranean (Cita); and to figure out the meaning of the M reflector and how the Mediterranean Basin had formed (Ryan). Early in 1970, the *Glomar Challenger* sailed out of Lisbon, Portugal, and through the Straits of Gibraltar, then immediately began drilling sites in the Balaeric Basin (the western half of the Mediterranean between Spain and Sardinia-Corsica, including the Balaeric Islands). Their first sites (Sites 120, 121, and 122) were near the western edge of the Balaeric Basin, just off the east coast of Spain (figure 24.7). After drilling through thick sequences of Pleistocene and Pliocene marine muds, they encountered a thick layer of gravel at the Miocene-Pliocene boundary, suggesting that coarse river sediments and even flash floods from arroyos, or dry desert washes, had flowed deep into the Mediterranean Basin. Although not conclusive, this strongly indicated that the upper slopes of the Balaeric Basin were not always under seawater but had once been broad alluvial fans covered with sand and gravels washing down from the land.

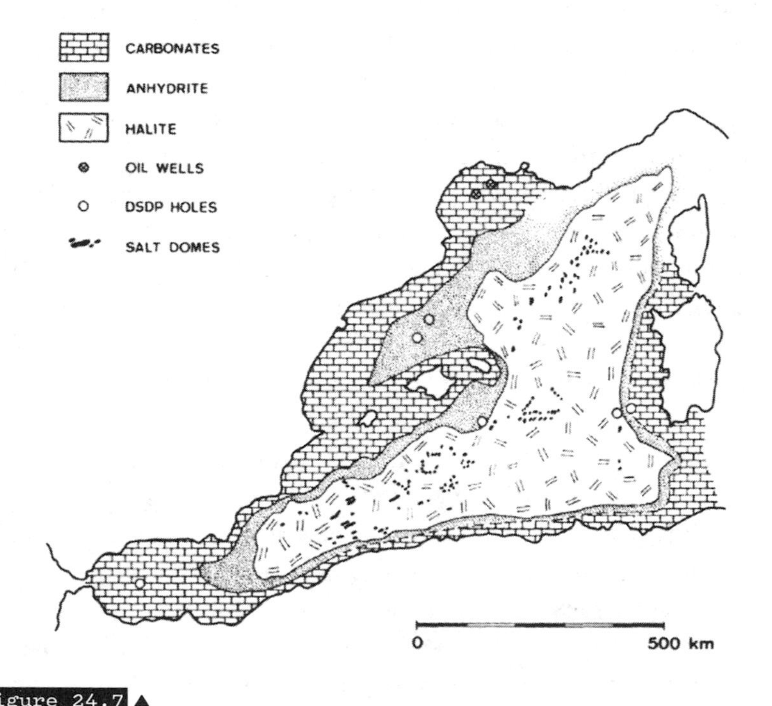

Figure 24.7 ▲

Concentric zonation of evaporite minerals in the Balaeric Basin of the western Mediterranean, showing the location of DSDP sites. (Courtesy of the Deep Sea Drilling Project)

Then they directed the ship to drill farther east and down the slope of the basin. Once they got below the Pliocene and Pleistocene muds, they found something really surprising: laminated stromatolites (see chapter 13) and crusts of dolomite, indicating that the lower slope was a super-salty intertidal mud flat, where only blue-green bacteria and some algae could form mats. The water had been so salty that odd minerals like dolomite were being formed. This strongly pointed to the bottom of the Mediterranean being exposed to drying, because stromatolites and intertidal deposits of dolomite do not grow in the deep ocean, and photosynthetic bacteria cannot live without sunlight.

Finally, they sent the ship to the center of the "bull's-eye"—the center of the Balaeric Basin. Sure enough, once they drilled below the Pliocene marine sediments, they found thick layers of salt and gypsum, thousands of feet below the present bottom of the Mediterranean Sea. This was rock-solid proof that the western Mediterranean had completely dried up about 5.5 Ma. And it also confirmed what many seismologists had suspected: the M reflector was a thick layer of uppermost Miocene salt and gypsum found beneath the bottom layers across the entire Mediterranean.

It was the same kind of deposit you find in any body of water that dries up completely, such as the Dead Sea or Death Valley. On the fringes of such basins would be alluvial gravels, and sometimes intertidal deposits with stromatolites and carbonate minerals in marine basins. A set of famous experiments by the Italian chemist M. J. Usiglio in 1849 had shown that about 50 percent of the original water must be evaporated for carbonates (like calcite, aragonite, or dolomite) to form. The next zone in from the outer ring of the bull's-eye is rich in sulfates like gypsum, which forms when the evaporation has removed 80 percent of the original volume of water. Finally, the last salts to be precipitated are the "bitter salts": halite (sodium chloride), sylvite (potassium chloride), plus different forms of calcium chloride. These only form when 90 percent or more of the original water has evaporated away; they precipitate out of the highly concentrated brine as the last of the water vanishes from the middle of the basin—the center of the bull's-eye.

Surprising discoveries kept on coming. In one core, they drilled through a windblown dune deposit from the desert at the bottom of the Mediterranean, made of grains of quartz sand mixed with the dried shells of plankton that had weathered out of the old marine muds, blown by an ancient

Miocene sandstorm. In others, they found abundant mud cracks, proof that the bottom had dried up completely. In most cores, they found layers of salt and gypsum alternating with layers of normal marine sediment, showing that the bottom of the Mediterranean had dried up, then briefly flooded again, then dried up again, over and over. The repeated episodes of drying and then reflooding is what allowed the salt and gypsum deposits in the Strait of Messina to reach such enormous thicknesses, because a single evaporation event would only leave one thin layer of salt and gypsum once all the water was gone. In fact, the amount of salts is estimated at 4 trillion trillion kilograms, or more than 1 million cubic kilometers! This amount of salt is 50 times what the Mediterranean now holds. Thus, it would require the Mediterranean to be dried up at least 50 consecutive times, so the rapid fluctuations between flooding and drying must have been dramatic.

SOLUTION: A GIANT DEAD SEA

DSDP Leg 13 continued on east past Italy and Sicily, drilled more holes in the eastern Mediterranean, and found more of the same kinds of deposits. By the time the voyage was done, the evidence seemed incontrovertible to the scientists on board. In their minds, there was just no other way to explain it. The Mediterranean had once been a huge desert (figure 24.8).

Yet as soon as they returned from the voyage and began presenting their results at scientific meetings in early 1971, they were met with incredible skepticism and resistance. No matter how strong the evidence, many scientists could not wrap their minds around the idea that the entire Mediterranean Sea had once dried up and left a desert basin up to 6,000 feet below sea level, filled with salt—a gigantic version of the Dead Sea. Although there are still scientists who resist the idea 47 years later, most have accepted the conclusions that Hsü, Ryan, and colleagues advocated in 1970 and 1971.

The first thing to realize is that Mediterranean is a very vulnerable body of water. It lies in the subtropical high pressure–belt desert latitudes (see figure 18.4), with regions that are mostly hot and dry, from the coast of North Africa to the European countries with their warm dry "Mediterranean climates," such as Spain, Italy, and Greece. As a consequence of lying in the belt of deserts, the region gets much less water flowing in than evaporates out, and only a few large rivers (the Nile and the Rhône) add fresh

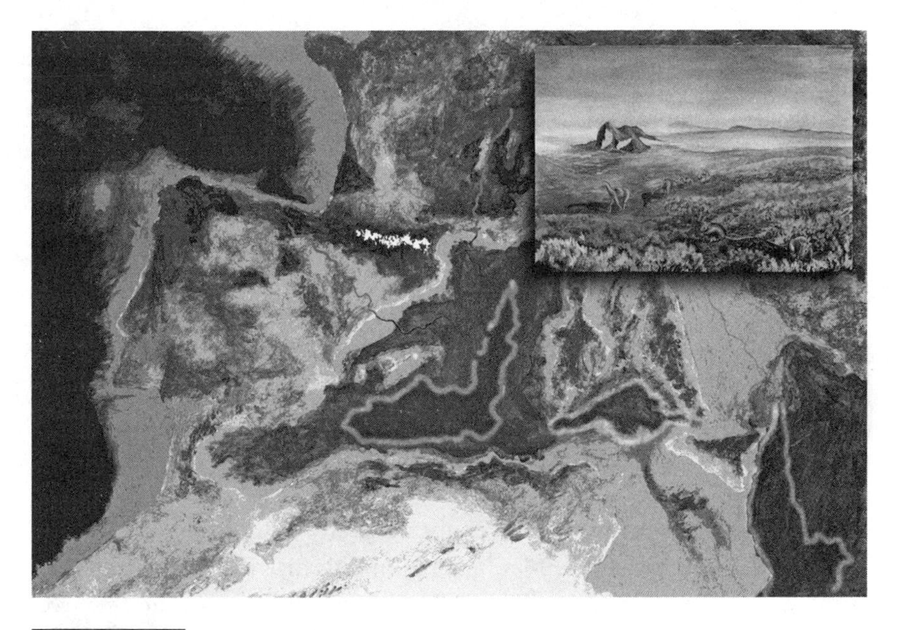

Figure 24.8 ▲
An artist's conception of the drying up of the Mediterranean Basin. (Courtesy of Wikimedia Commons)

water to replace this high rate of evaporation. The basin was first closed at its eastern end when the Arabian Peninsula and Africa collided with Asia about 20 Ma. That leaves only the narrow Straits of Gibraltar to allow seawater to rush in and replace all the evaporated water. All it would take would be one critical event to shut this narrow spigot, and the water would be cut off completely.

Indeed, at 5.96 ± 0.02 Ma, there was a global drop in sea level at the same time the collision between Africa and Spain was causing the Atlas Mountains to rise up. Once sea level dropped below the narrow sill that allowed it to flow in, the Mediterranean was cut off. Between 5.96 and 5.33 Ma, the Mediterranean Basin was alternately dried out completely, then flooded when water leaked through Gibraltar for a while, then flooded again, and again, at least 50 different times.

Meanwhile, the basin would have been like an enormous Dead Sea, and animals that could tolerate the heat and dryness could migrate across the Mediterranean or between many areas that are now islands. In fact, we have long known that many of the islands of the Mediterranean—Cyprus,

Malta, Gargano, Corsica, Sicily, and Sardinia—have unusual dwarfed fossil mammals that evolved in isolation once these regions were cut off when the Mediterranean flooded for the last time. Cyprus and Malta each had their own extinct dwarfed hippos. Other islands have dwarfed mammoths. Still others have gigantic dog-sized hedgehogs, and rabbits the size of hogs. This is common in many island faunas, where large mammals such as elephants or hippos become smaller because of lack of predators and limited resources, while small mammals like hedgehogs and rabbits become large because of the absence of other mammals that would normally compete with them in that body-size class.

In fact, mammalian paleontologists had long known that there was a major change in the mammalian faunas at the end of the Miocene, with mammals crossing freely between Eurasia and Africa. Based on this evidence, they were the first to postulate that the Mediterranean had dried up.

The realization that the Mediterranean had dried up also solves the final piece of the puzzle: the Grand Canyon of the Nile. As the sea level of the Mediterranean began to drop early in the Messinian event, the ancient predecessor of the Nile River would have cut down into its floodplain and canyon to match its gradient. This would continue until the Mediterranean had completely vanished, at which time the Nile Canyon would be cut down all the way to about 2,500 meters (8,000 feet) below sea level to flow into the salt pans, then evaporate. When the flooding refilled the basin in the Pliocene, the Grand Canyon of the Nile would have been full of seawater, and soon filled in with both marine sediments from the ocean and sands and muds from the Nile headwaters upriver. This idea was confirmed when studies were made of the Rhône Valley in France, which also has a deep canyon beneath it now completely filled with Pliocene marine sediments.

Finally, the fact that 50 smaller episodes of flooding just evaporated away gives us some idea of how much water was needed and how fast it must flow to compensate for the high rate of evaporation. Indeed, an even larger volume of water would be necessary for the great Pliocene "deluge" that finally filled the Mediterranean and ended the great cycle of drying and flooding. Ryan calculated that there needed to be so much cold Atlantic seawater rushing through Gibraltar so fast that it would have made a waterfall about 1,000 times as big as Niagara Falls, or 15 times as big as the Zambezi Falls, one of the largest in the world. Ryan and Hsü calculated that about

34,000 cubic kilometers of water would have to flow through the Gibraltar Falls in a year to fill up the basin in 100 years, and it was probably much more. At those rates, the volume and pressure were so great that the roar would have broken the sound barrier!

From a simple observation of salt in the Strait of Messina and the canyon beneath the Nile, our understanding of one of the most amazing episodes in the history of the earth emerged.

FOR FURTHER READING

Bascom, Willard. *A Hole in the Bottom of the Sea: The Story of the Mohole Project.* New York: Doubleday, 1961.

Briggs, Peter. *200,000,000 Years Beneath the Sea: The Story of the* Glomar Challenger*—The Ship that Unlocked the Secrets of the Oceans and Their Continents.* New York: Holt, 1971.

Hsü, Kenneth J. Challenger *at Sea: A Ship That Revolutionized Earth Science.* Princeton, N.J.: Princeton University Press, 1992.

——. *The Mediterranean Was a Desert: A Voyage of the* Glomar Challenger. Princeton, N.J.: Princeton University Press, 1983.

25 A POET, A PROFESSOR, A POLITICIAN, A JANITOR, AND THE DISCOVERY OF THE ICE AGES

GLACIAL ERRATICS

The glacier was God's great plough set at work ages ago to grind, furrow, and knead over, as it were, the surface of the earth.

—LOUIS AGASSIZ

PUZZLE #1: WANDERING BOULDERS

In the early days of geology in the late 1700s and early 1800s, one of the greatest puzzles was the occurrence of huge boulders, often found in odd places or balanced in precarious positions (figure 25.1). Even more puzzling was the nature of these rocks. They were formed of rock types that came from nowhere near the place where they now resided. In some cases, the origin of these rocks could be traced to unique source areas hundreds of miles to the north. The pioneering Scottish geologist (and protégé of James Hutton) Archibald Geikie much later described them as "large masses of rock, often as big as a house, that have been transported by glacier-ice, and have been lodged in a prominent position in the glacier valleys or have been scattered over hills and plains. An examination of their mineralogical character leads to the identification of their sources."

The more geologists looked at them, the more they discovered how incongruous they were. In Schokland in the middle of Holland, a country covered by very young sediments with not a single outcrop of hard bedrock anywhere, stands a huge set of boulders that can be traced to Norway.

Figure 25.1 ▲

Several examples of huge precariously balanced erratic boulders in northern Europe: (*A*) *Kummakivi* ("strange rock" in Finnish), in Savonia, Finland; (*B*) a giant out-of-place boulder perched on other rocks in Norber, near Austwick, Yorkshire, England; (*C*) an erratic balanced boulder recently dropped by the melting of glaciers in Yosemite National Park, California. (Courtesy of Wikimedia Commons)

Figure 25.1 ▲
(continued)

In the middle of the north German coastal plain near Stöckse sits another set of huge boulders, called the Giebichenstein. They are clearly not local; the region has no hard bedrock, just soft coastal plain sediments. These rocks can be traced to Scandinavia. There were many more in Europe, and even more remarkable ones on other continents, such as the huge boulders sitting on top of the plains of Alberta, clearly not derived from the soft Cretaceous shales that surround them. Plymouth Rock, on which the Pilgrims landed in 1620, is another example. When I used to run geology field trips in central Illinois, we would often stop at a certain roadcut and pull out lumps of native copper, which can only have originated from the Upper Peninsula of Michigan. Examples like these were found all around the Northern Hemisphere, and became known as "wanderers" or "erratics," from the Latin verb *errare*, "to wander." (Words like "err" and "error" come from the same root—when you err, you wander from the truth. If you walk "erratically," you are wandering.) They were also known as "lost sheep" or "foundlings," since they reminded scholars of sheep that had strayed far from their home flock.

As we saw in chapter 5, however, most early geologists attributed all layered rocks (not just sediments, but even lava flows) to the actions of a great flood, usually considered to be Noah's flood. This might explain the long-distance displacement of the rocks but neglected important details, such as how water could pick up such large boulders and not leave entire beds of gravel and cobbles (as happens now with powerful floods). It also did not explain the fact that many of these boulders are very sharp and angular, not well rounded, as always happens when boulders and cobbles are rolled and tossed along in water, especially in flash floods. But geology was still in its infancy at this point, and the Great Flood dogma was so predominant that few people were looking for problems with it. Instead, they ignored all sorts of inconsistencies in order to shoehorn everything they found into this one simple explanation. In 1824 the prominent British naturalist William Buckland wrote:

> We have evidence, that a current from the north has drifted to their present place, along the whole east coast of England, that portion of the pebbles there occurring, which cannot have been derived from this country; a certain number of them may possibly have come from the coast of Scotland, but the greater part have apparently been drifted from the other side of the German ocean.

Thus, the glacial erratics were a mystery. So too were the thick deposits of unsorted boulders and sand and clay that early geologists called the "Drift," as in deposits that had formed by the drifting of flood currents. Another name for it was the "Diluvium," literally the "flood deposits." Once again, if they had looked critically, they would have seen that any current of flowing water generates distinct layers and bedding, no matter how energetic the flood. Even when floods suddenly lose energy, they don't drop their load in a random mixture of unstratified boulders, gravel, sand, and clay, but deposit material sorted by layers of different grain sizes, from discrete gravel beds to sand layers to finely laminated deposits of silt and clay. But such critical thinking was still in the future . . .

PUZZLE #2: SCRATCHES IN ROCK

Yet another feature of the bedrock geology that puzzled early European geologists was the presence of long parallel scratches and grooves,

Figure 25.2 ▲
Glacial scratches in bedrock, caused by rocks in the base of the glacier being dragged across the bedrock like teeth in a rasp. (Courtesy of Wikimedia Commons)

sometimes running for many meters, and carved into very hard bedrock (figure 25.2). Sometimes the grooves were very shallow, but often they were incredibly deep. In some cases, there were huge areas of bedrock with parallel scratches gouged into them. What force could scrape grooves into the hardest bedrock like this? And more importantly, what caused them to all line up so highly parallel to one another?

Up until the late 1830s, the Great Flood again served as a convenient and simple idea to explain them. In 1824, Buckland wrote a 200-page monograph, *Reliquiae Diluviae* [Latin for "remains of the flood"]; *or, Observations on the Organic Remains attesting the Action of a Universal Deluge*, outlining the Noah's flood explanation for the erratics, the glacial drift, and the parallel scratches. He wrote that they were scoured "by the attrition of heavy bodies, set in motion by a great force of water in rapid movement." Yet any astute observer, then or now, if he or she watched the action of churning floodwaters, would realize that running water does not carry rocks in long straight lines for long distances and create gouges with them. Such experiments were in the future, however, since the geologists of that time were satisfied with the Great Flood model of the earth.

SOLUTION: AGASSIZ AND THE *EISZEIT*

Not all Europeans saw these strange geologic features as products of floodwaters. Those who lived near the Alps and their glaciers, in particular, had a different perspective, thanks to seeing glaciers in action. As early as 1787, Swiss minister Bernard Friedrich Kuhn argued that erratic boulders were carried by glaciers, because he could see Swiss glaciers carrying such boulders. When James Hutton visited the Jura Mountains of Switzerland and France a few years later, he reached the same conclusion—something that was overlooked among all his radical and revolutionary ideas. In 1824 the Norwegian naturalist Jens Esmark argued that glaciers had caused the scratches and erratics in Norway. They could see how the immense weight of the ice pushed down on the rocks being dragged along the bottom of the glacier and put enormous force on them, scoring the bedrock like teeth on a rasp. Influenced by Esmark, German naturalist Reinhard Bernhardi published an article in 1832 arguing that there was once a polar ice cap across all of Europe, even to central Germany.

Meanwhile, in Switzerland the observations of modern alpine glaciers and their effects were accumulating and becoming better understood. In 1815 the Swiss mountaineer and chamois hunter Jean-Pierre Perraudin described the effects of glaciers on the Swiss valleys and inferred that the glaciers had once been larger and extended farther. By 1818 his ideas had impressed a highway engineer by the name of Ignace Venetz, who had spent much time in the Swiss landscape as part of his work. Venetz gradually became more convinced that glaciers had spread out from the Alps and affected the surrounding area. In presentations given in 1816 and 1821 and finally in 1829, he committed to the idea of a great expansion of glaciers in the past. Meanwhile, the director of the Bex salt mines, Jean de Charpentier, also became convinced after listening to Perraudin and then to Venetz and doing his own field observations. He published several more-complete articles on the topic from 1829 to 1833.

In the audience of de Charpentier's presentation in Lucerne in 1834 was a promising young Swiss paleontologist by the name of Louis

Figure 25.3 ▲
Louis Agassiz as a young man. (Courtesy of Wikimedia Commons)

Agassiz (figure 25.3). He was already famous through much of Europe for his pioneering work on fossil fish. In the summer of 1836, he visited de Charpentier in Bex, with the intention of proving his glacial ideas wrong. Instead, he returned a convert to the glacial theory and was eager to spread this idea among geologists outside the handful who knew glaciers well. Unlike earlier advocates of glacial geology who simply gave evidence for glaciation of the Alps and regions nearby, Agassiz was an imaginative thinker, bold speaker, provocative writer, and an eager, hardworking, and ambitious man as well.

In 1837 Agassiz was hosting the annual meeting of the Swiss Society of Natural Sciences at his home base of Neuchâtel. When he rose to give the opening address, the audience expected another presentation on fossil fishes. Instead, he launched into a radical argument in favor of the glaciers as causes of most of the features once attributed to Noah's flood. After reviewing the evidence and the ideas of Perraudin, Venetz, and de Charpentier, he extended their thinking to most of Europe and argued that it had once been covered by ice during an "Ice Age" (*Eiszeit*). His presentation was so unexpected and surprising that the meeting was thrown into confusion, and the long debate completely derailed the program of other talks that were supposed to be given. One of these was by Amanz Gressly, who was planning to introduce his now-famous concept of "sedimentary facies" but never had a chance to give his presentation during the chaos that followed Agassiz's bombshell.

Most of the audience was still highly skeptical and critical of Agassiz's ideas, so he responded by throwing together a spur-of-the-moment field trip to the nearest alpine glaciers at the end of the meeting. (This would be impossible at a professional scientific meeting today, where everything is tightly scheduled months in advance, and people have planes to catch and hotel reservations that can't be changed.) The greatest geologists of the region in attendance, including Élie de Beaumont and Leopold von Buch, rode in the carriage with him. If Agassiz thought they would be instant converts upon seeing the field evidence, he was unduly optimistic about human nature. They ended the field trip unconvinced, and most of the geological community was still critical of his idea. The great naturalist and explorer Alexander von Humboldt told Agassiz to go back to fossil fishes and "render a greater service to positive geology, than by these general considerations (a little icy besides) on the revolutions of the primitive world, considerations which, as you well know, convince only those who give them birth."

If scientists on the Continent gave Agassiz a cold reception, others were more receptive. In England, William Buckland heard about some of his ideas and began to have doubts about his own work explaining everything with Noah's flood. Buckland had hosted Agassiz in 1835 in Oxford while the latter was visiting and studying fossil fish, and they had become friends. In 1838 Buckland went to Freiburg, Germany, to attend the Association of German Naturalists and listen to Agassiz promote his glacial theory and give it a fair hearing. Then he went with Agassiz to Neuchâtel to get a grand tour of glacial geology, accompanied by Charles-Lucien Bonaparte, a wealthy patron of the natural sciences. (Bonaparte had little else to do with his time since his cousin lost the Battle of Waterloo in 1815 and went into his final exile.)

Buckland got the grand tour, but left without being totally convinced. He ruminated over what he had seen until 1840, when Agassiz addressed the British Association of the Advancement of Science in Glasgow. Finally, Buckland became a convert and was one of the first Britons to become a glacial believer. Most of the rest of the British geologists were surprised and scornful of his abrupt about-face on the issue. A famous satirical cartoon of the time (figure 25.4) shows Buckland clothed in his normal field attire of top hat and academic gown, carrying charts, hammers, and other geological equipment, and standing directly on a surface with parallel scratches. The scratches are labeled "Scratched by a glacier thirty three thousand three hundred and thirty years before the creation," and "Scratched by a cart wheel on Waterloo Bridge the day before yesterday."

Buckland then convinced the most influential critic of all, Charles Lyell, who immediately wrote a paper in support of his ideas. Then Agassiz, Buckland, and Lyell toured the Scottish Highlands, where Agassiz was able to show his colleagues that many of the distinctive features that had so long remained unexplained were due to the fact that Scotland had once been overrun with glaciers. By 1841 Edward Forbes was writing to Agassiz, "You have made all the geologists glacier-mad here, and they are turning Great Britain into an ice house. Some amusing and very absurd attempts at opposition to your views have been made by one or two pseudogeologists."

The battle over the Ice Ages raged on for many years more, but Agassiz had grown tired of the continuous unresolved conflict. He sailed to the United States in 1846, originally to study fossil fish, but then received a very generous offer to stay on at Harvard and found the Museum of Comparative Zoology. He remained in the United States for the rest of his life, dying in 1873 after 27 years of teaching and research at Harvard. More importantly, he used the change in venue to take field trips all over the northeastern part of North America, where he found one glacial feature after another, further confirming his idea that it was indeed a global expansion of the Arctic ice cap that had covered all the northern continents. He also trained a number of prominent students who formed the next generation of American zoologists and geologists, including paleontologists Charles Doolittle Walcott (chapter 17), Alpheus Hyatt, and Nathaniel Shaler; paleontologist and entomologist Alpheus Packard; pioneering naturalist Ernest Ingersoll; ichthyologist David Starr Jordan; explorer-geologist Joseph LeConte; and the famous philosopher and psychologist William James.

Figure 25.4 ▲

Famous satirical cartoon of William Buckland standing on glacial scratches. (Courtesy of Wikimedia Commons)

A professor, a politician, and a poet. The Swiss professor Agassiz had launched the glacial Ice Age theory, and the politician-geologist Lyell had thrown his support behind it, convincing many other geologists by 1842. But it would take another discovery by a poet-explorer named Elisha Kent Kane to finally convince the world.

PERIL AND DEATH IN GREENLAND

One the main problems for geologists and other people trying to wrap their minds around the concept of an "Ice Age" was visualizing huge ice sheets covering all of Europe. It was one thing to think of the alpine glaciers extending much farther in the past than today. It was another to imagine the thick polar ice cap covering all of Europe. Once again, the problem lay with the limited knowledge that most Europeans had of the world. Just as Wernerians had never seen a volcano erupt and so could not visualize a lava flow as liquid rock, so too the geologists of the time had no idea what a giant ice sheet might look like.

Up until then, nearly all the breakthroughs in science and scholarship were the domain of Europeans. But the young nation of the United States began to invest in exploration and expansion, with pioneers and scouts who first traveled from the Appalachians to the Mississippi, and in the Lewis and Clark expedition in 1803–1805, from the Mississippi all the way to the Pacific. In 1848 the United States had won huge territories in the Mexican-American War, and by 1850 California was admitted to the Union as a state.

Americans were exploring their western lands, but there was also a push to discover the northern polar regions as well. A number of daring Americans tried to make their mark by seizing this new opportunity. One of these was a young man named Elisha Kent Kane. Born of a prominent Philadelphia family in 1820, he received his medical degree from the University of Pennsylvania when he was 22, then became an assistant surgeon in the U.S. Navy. This placed him in the midst of many perilous missions: the China Commercial Treaty mission, a mission to Africa, and several battles in the Mexican-American War. At the Battle of Nopalucan on January 6, 1848, Kane captured, then made friends with, Mexican general Antonio Gaona and his wounded son.

In 1845 British explorer and naval officer Sir John Franklin had embarked on a daring voyage to find the Northwest Passage through the Arctic, using two sailing ships, HMS *Erebus* and HMS *Terror*. The expedition had sailed north and then vanished, with no reports of what had happened to them. As Franklin was a prominent British aristocrat and explorer, there was a huge reward from Lady Jane Franklin and the British Admiralty to find Franklin and his ships and crew. Many expeditions then set out to find the missing men and claim the reward, so that by the summer 1850, there were 11 British ships and two American ships searching for the Franklin expedition.

Kane's experience and daring led him to be appointed as the senior medical officer of an expedition financed by Henry Grinnell in 1850 to find the lost expedition of Sir John Franklin. They were the only such expedition to be moderately successful, since they found the location of Franklin's first winter camp. Buoyed by this success, Kane organized a second expedition financed by Grinnell to search farther. They sailed from New York on May 31, 1853, and reached Rensselaer Bay in Greenland by winter. Through the winter he and his crew suffered from scurvy, and some were near death. Nevertheless, the next spring Kane's group kept forcing their way north, where they discovered the ice-free Kennedy Channel in the northernmost gap between Greenland and Ellesmere Island. This became the preferred route for future polar explorers, culminating with the discovery of the North Pole by Robert Peary in 1910.

However, their travels through Greenland were getting more arduous, and soon their task was not to find Franklin, but to survive and get home themselves. By May 20, 1855, their brig, the *Advance*, had become icebound, and they could no longer sail. Kane then led his crew on a march across the ice sheet for 83 arduous days until they finally reached an open-water passage near Upernavik (halfway up the west coast of Greenland), where they were rescued by a passing sailing ship. Despite the hardships, all but one of the men endured the perilous march across the treacherous crevasses and survived sub-freezing conditions, even though they were hampered by the weight of the ice sledges, dragging their invalids with them.

On October 11, 1855, the Kane expedition finally returned to New York and received a hero's welcome. Kane then set about writing up his notes of the expedition, and a year later the two-volume report was published. It was a sensation in both the United States and Europe, because it was a gripping account of perilous and heroic deeds that galvanized the readers. In addition, this was the first time anyone had survived and returned to describe the nature of an ice sheet a mile thick. His report became a runaway best seller, and both scientists and layman alike were able to visualize and understand the concept of immense sheets of ice for the first time. Within a few years, the remaining geologists who objected to Agassiz's *Eiszeit* were no longer able to dismiss the concept of a giant ice sheet. Geologists could finally imagine miles of ice piled on top of Europe and North America.

Two final notes on this story: Kane himself was still ill and recovering from the extreme stress of the expedition, but he felt obligated to sail to England and deliver his report to Lady Franklin himself. On doctor's orders, he then sailed to Cuba to recuperate from his exertions. Unfortunately, his health only got worse, and he died in Cuba on February 16, 1857. He was only 36 years old. His body was brought back to New Orleans, then a train carried it home to Philadelphia. At stops all along the way, the train was met by a memorial delegation honoring their national hero. It was the longest funeral train in American history up to that time, eventually exceeded by the funeral train that carried Abraham Lincoln home to Illinois after he was assassinated in April 1865.

Finally, none of those early expeditions found out what happened to the ill-fated Franklin expedition. Only with the advent of modern satellite technology and superior methods of searching the Arctic have they finally been found and their fate determined. We now that the entire expedition of 129 men was icebound, and all died of disease, hypothermia, starvation, or cannibalism. The wreck of HMS *Erebus* was found in 2014, and HMS *Terror* was only found in 2016. Both had become icebound, then crushed by ice, and sank in deep water.

THE SCOTTISH JANITOR AND THE SERBIAN MATHEMATICIAN

After the confirmation of the Ice Age idea, geologists set to work through the late nineteenth and much of the twentieth century mapping the glacial "drift" deposits (actually moraines made of glacial till, formed when sand, gravel, and boulders pile up at the snout of a melting glacier). Soon they were able to establish that there had been four huge ice advances in North America, named (from youngest to oldest) the Wisconsinan, Illinoian, Kansan, and Nebraskan Stages. Their names came from the southernmost state in which they left moraines. Meanwhile, European geologists had found evidence of five successive glacial advances, named (youngest to oldest)

Würm, Riss, Gunz, Mindel, and Donau. There was not just one Ice Age but at least four or five of them on the northern continents. Unfortunately, there were no dating methods to determine how old these glacial advances were or to determine how the four events in North America might match up with the five events in Europe. Geologists put out all sorts of explanations for the evidence of multiple glacial advances, but none were very well supported.

One idea that was being discussed among astronomers was that the amount of sunlight received on the earth's surface (known as solar insolation) might determine the shift from glacial ice advances to interglacial ice retreats. No one had really worked out the mathematics of the required orbits or calculated the energy differences involved. Into this breach stepped a remarkable Scotsman by the name of James Croll (figure 25.5). He was the

Figure 25.5 ▲
James Croll. (Courtesy of Wikimedia Commons)

perfect example of someone who rose from limited means and education and reached the pinnacle of his profession through sheer effort and intelligence. Born on a farm in Perthshire, Scotland, he had almost no formal schooling and had to go out to work before he was 16 years old. He started as an apprentice wheelwright and mechanic, then became a tea merchant, then failed after trying to run a temperance hotel, and then became an insurance agent. In 1859 he was hired as the janitor at the Andersonian University in Glasgow, where he used his access to spend many hours in the library, teaching himself mathematics, physics, and astronomy.

Building on the work of astronomer Urbain Le Verrier, who first showed that the earth's orbit around the sun and axial tilt were constantly changing, Croll showed his work to Sir Charles Lyell and Sir Archibald Geikie, who were impressed. Geikie hired him as keeper of the geologic maps and correspondence of the Geological Survey of Scotland in Edinburgh, where he had plenty of spare time to read and do his research, surrounded as he was by all the necessary documents. In 1875 he wrote the book *Climate and Time, in their Geologic Relations*, which laid out the basics of how the earth's orbital motions change the amount of solar radiation received, and thus trigger Ice Ages. This earned him a university research post, an honorary degree from the University of Saint Andrews, and eventually election to the Royal Society of London.

Croll did calculations for the known cycles of the earth's orbit around the sun, to see how much they might explain the Ice Ages. He pointed out that astronomers as early as Johannes Kepler in 1609 knew the earth's orbit around the sun was not a circle but an ellipse. That ellipse changed shape from nearly circular to more egg-shaped very slowly (we now know it takes about 100,000 years). This is the cycle of the "eccentricity" of the earth's orbit (figure 25.6A). Another cycle, known since the days of Hipparchos in 130 B.C., was the precession or "wobble" cycle. As the ancient Greeks knew, the earth's axis wobbles like a top, with its spin axis pointing in different directions. Today, for example it points to Polaris, the North Star, but 10,000 years ago it pointed to a completely different star, Vega. As it points in different directions, it affects the amount of sunlight the poles receive. This is the precession or "wobble" cycle, which we now know takes about 21,000–23,000 years to complete, the fastest of the three cycles.

Croll also pointed out that the ice can grow and melt back very rapidly due to the albedo feedback loop. When there is a lot of ice, it is very

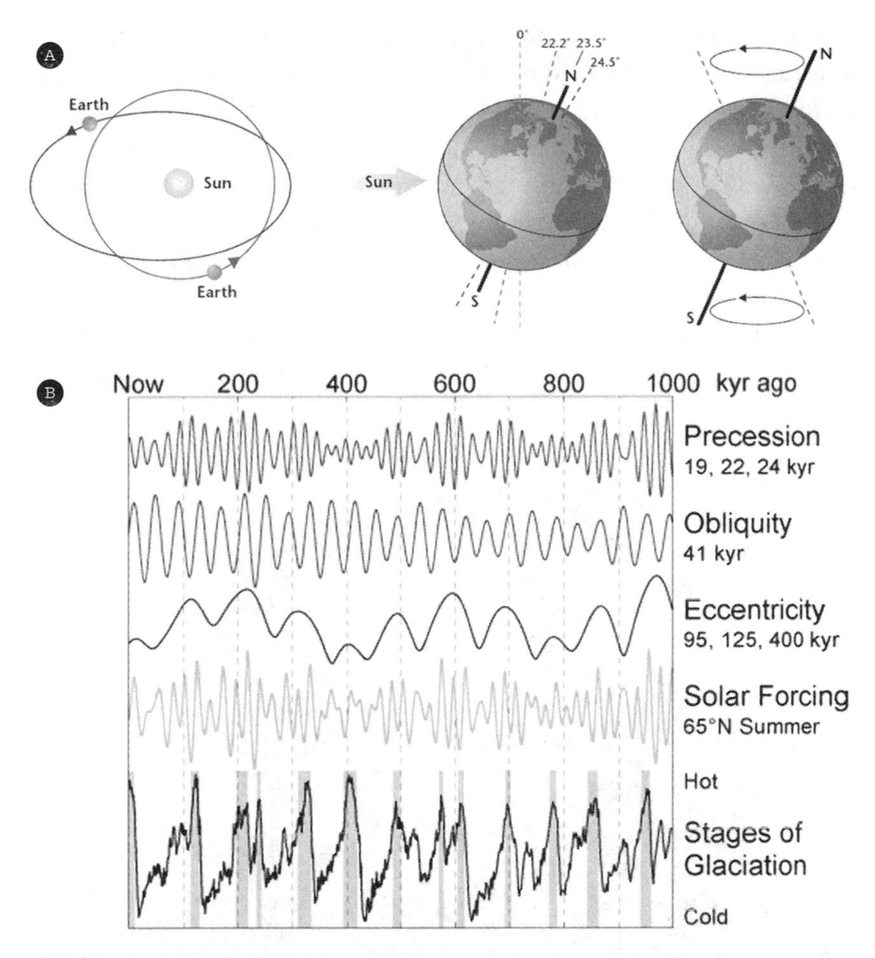

Figure 25.6 ▲

Milankovitch cycles: (*A*) A diagram showing the three cycles of changes in the earth's orbit and angle toward the sun. (*B*) Each cycle has its own periodicity, forming three sine waves that interact with one another to form a complex interference pattern, which produces a saw-toothed pattern of warming and cooling. This pattern was confirmed by measurements of the ancient ocean temperatures recorded in deep-sea cores, and eventually in bubbles of ancient air trapped in ice cores. (Courtesy of Wikimedia Commons)

reflective, or has a high albedo. This tends to bounce back more solar energy to space and cool the temperature, further increasing the ice. But if you have an ice-covered landscape, all it takes is a little melting to expose dark, sunlight-absorbing, low-albedo surfaces like seawater or vegetation, which in turn absorb more heat and accelerate the melting.

Croll's book was very provocative and worth taking seriously, but unfortunately, there were no data suitable to test his ideas at that time. Nothing could be dated reliably back then, and the land record of ice advances was too incomplete and poor to evaluate his ideas. So they languished in the pile of interesting but untestable ideas for decades. Croll himself sustained a serious head injury in 1880 that forced him to retire at age 59, and he lived 10 more years before dying in 1890, with no further progress made on his pioneering suggestions.

Croll's ideas were nearly forgotten when a Serbian astronomer and mathematician named Milutin Milankovitch revived them (figure 25.7).

Figure 25.7 ▲

Milutin Milankovitch. (Courtesy of Wikimedia Commons)

Born in what is now Croatia (part of the Austro-Hungarian Empire back then) in 1879, he was a stellar student, and received an engineering degree from the Vienna Institute of Technology in 1904. He became a top civil engineer, building bridges, viaducts, aqueducts and dams, and other structures in Austria. He even received six patents for his inventions, and then moved to Serbia to become chair of applied mathematics at the University of Belgrade.

Despite his day job as an engineer, he was more interested in fundamental research, and by 1912 he was involved in solving the problems of how variations in solar insolation might affect climate. He lamented that "most of meteorology is nothing but a collection of innumerable empirical findings, mainly numerical data, with traces of physics used to explain some of them . . . Mathematics was even less applied, nothing more than elementary calculus . . . Advanced mathematics had no role in that science." In 1912 and 1913, he published several papers calculating the amount of solar insolation that the earth receives at each latitude, and how this affects the position of climatic belts. Then, in July 1914, Archduke Franz Ferdinand was assassinated in Sarajevo, and the crisis between Serbia and the Austro-Hungarian Empire boiled over to become World War I. As a Serbian, Milankovitch was arrested during his honeymoon in Austria, then he was imprisoned in Esseg Fortress. About his first night as a POW, he wrote:

> The heavy iron door closed behind me. . . . I sat on my bed, looked around the room and started to take in my new social circumstances . . . In my hand luggage which I brought with me were my already printed or only started works on my cosmic problem; there was even some blank paper. I looked over my works, took my faithful ink pen and started to write and calculate . . . When after midnight I looked around in the room, I needed some time to realize where I was. The small room seemed to me like an accommodation for one night during my voyage in the Universe.

Fortunately, he had powerful connections in Vienna, so he was then interned in Budapest, with access to materials so he could continue his research. Even though he was technically a prisoner, he used his undisturbed time in Budapest and his university library access to make huge advances in mathematical meteorology, which he published in a series of papers from 1914 to 1920. Finally, the war ended, and Milankovitch and his family returned to Belgrade in March 1919, where he resumed his professorship at the University of Belgrade.

Then Milankovitch built on this research to calculate exact models for how solar insolation cycles might cause Ice Ages. Crucially, he realized that the key factor is how much sunlight the earth's surface receives in the summer time, which determines how much the ice melts back—or remains. Milankovitch built on previous understanding of Croll's eccentricity cycle and the precession cycle, and added a third that had been discovered by Ludwig Pilgrim in 1904: the obliquity or "tilt" cycle. The rotational axis of the earth is not straight up-and-down with respect to the plane of its motion around the sun, but tilted at 23.5° (see figure 25.6A). That angle is not always constant, but fluctuates between about 22° and more than 24.5°. When the angle is as steep as 24.5°, the polar regions get a lot more sunlight and the ice melts; when it is as shallow as 22°, the poles get much less sunlight and ice forms. This is known as the obliquity cycle of the earth's orbit and takes about 41,000 years to run in a complete cycle from 22° to 24.5° and back again. Milankovitch had all the pieces needed to do the painstaking calculations and plots (by hand on many reams of paper, with no computers or calculators). After dozens of scientific papers and short books on the topic of earth's solar radiation and climate, by the late 1930s Milankovitch focused on putting it all together in one book, *Canon of Insolation of the Earth and Its Application to the Problem of the Ice Ages*.

Once again, world events reached a crisis and interfered with Milankovitch's life and work. Four days after he sent the book to the printer in 1941, the Germans invaded Yugoslavia and the printing house was destroyed during the bombing of Belgrade. Luckily, the printed pages were in another warehouse and were undamaged and eventually bound and published. As the Nazis invaded Serbia in May 1941, two German officers and some geology students came to his home to help him, and he gave them his only bound copy of the book for safekeeping in case something should happen to him or his work. He spent the rest of the war years holed up in his home, writing his memoirs. When the war ended, he once again returned to his duties at the University of Belgrade and also served as vice president of the Serbian Academy of Sciences. Even though he had received many honors and awards, Milankovitch did not rest on his laurels, but continued to work on important problems. He took interest in revising the Julian calendar, examining the idea of polar wandering, and writing about the history of science, even after he retired in 1954. He suffered a stroke in 1958 and died at the age of 79, without ever knowing whether his ideas would be supported by geological evidence.

SOLUTION: PLANKTON AND THE PACEMAKER OF THE ICE AGES

But even though Milankovitch had brought the problem of the astronomical cycles and the causes of the Ice Ages as far as any astronomer or mathematician could do by calculation, there was still no clear evidence from geology to support it. The deposits on land still only showed four or five major glacial advances, and even in the late 1950s, their dates were still problematic. The entire idea of astronomical cycles and Ice Ages was still an unconfirmed speculation.

This problem could never be solved with the record of glaciations on land, because most of the land record is prone to being eroded away. It is full of gaps that make it incomplete. The solution finally came when long cores of deep-sea sediment were analyzed in the early 1970s. Unlike the incomplete record of sediments on land, the deep-sea floor is blanketed by an almost steady uninterrupted "rain" of fine muds and shells of plankton settling from the sea surface. Once there were enough good cores that had an almost continuous unbroken record of the last 2 to 3 million years of climate in the world's oceans, there was enough information to test the Croll-Milankovitch hypothesis.

Leading the research on this problem was a group of scientists funded by the NSF Project CLIMAP (an acronym for "Climate: Long-Range Investigation, Mapping, and Prediction"). The chief scientists were Lamont micropaleontologist James Hays (who was on my dissertation committee and also my coauthor on some of my early micropaleontology research), Brown University micropaleontologist John Imbrie, and Cambridge University isotope geochemist Nick Shackleton. The three of them analyzed many different deep-sea cores that had long continuous records of the entire last 2 to 3 million years of the Ice Ages. The cores were all precisely dated by the biostratigraphy of the microfossils, by volcanic ashes, and also by the flip-flops of the magnetic field recorded in the core sediments. The scientists found that certain temperature-sensitive plankton could be used to track the temperature changes in the ocean in any particular core. In addition, the chemistry of the minerals in the shells of the plankton can be used as a proxy for the temperature changes in the seawater, so there were several indicators of climate in these cores.

Once enough cores had been analyzed and correlated from all the world's oceans, the CLIMAP scientists discovered there were not just 4 or 5 ice age cycles, but more than 20 in the past 2 million years! Apparently, only the very biggest cycles that cause the largest glacial advances are recorded on land, but the traces of the smaller cycles are wiped out by the larger glacial advances that follow them. But in the deep-sea record, all the cycles are preserved, and the exact timing and magnitude of temperature change in the ocean can be plotted as a precise curve of temperature.

Once the CLIMAP scientists got their temperature curve, they tried to tease apart what might cause the complicated, sawblade-like like pattern of warming and cooling. Using a method called spectral analysis, the complex curve was analyzed and broken down into its components. It turned out to be a composite of three different sine waves that formed the complex interference pattern of the real data (figure 25.6B). Sure enough, the three frequencies that were behind the real data were the 110,000-year eccentricity cycle, the 41,000-year tilt cycle, and the 21,000- to 23,000-year precession cycle—just as Milankovitch had predicted over 30 years before.

And so, in 1975, exactly 100 years after Croll had published the first book on the topic, the problem was solved. In 1976 Hays, Imbrie, and Shackleton published their legendary "pacemaker" paper, which laid out all the evidence that astronomical cycles of the earth's motion around the sun dictate how much solar radiation we receive, and those cycles are the major controllers or "pacemakers" of the Ice Ages. Since that time, Croll-Milankovitch cycles have been discovered in all sorts of geological records, including the cycles of Carboniferous coal deposits and the chalk seas of the Cretaceous. The confirmation of the Croll-Milankovitch hypothesis has been considered one of the landmark discoveries in geology, and the Hays, Imbrie, and Shackleton "pacemaker" paper of 1976 is ranked as one of the most important scientific breakthroughs of the twentieth century.

And to think that the puzzle started with some huge boulders and scratched bedrock, but ended up being solved by the tiny shells of plankton in the deepest ocean.

FOR FURTHER READING

Gribbin, John, and Mary Gribbin. *Ice Age: The Theory That Came in from the Cold!* New York: Barnes and Noble, 2002.

Imbrie, John, and Katherine Palmer Imbrie. *Ice Ages: Solving the Mystery*. Cambridge: Harvard University Press, 1979.

Macdougall, Doug. *Frozen Earth: The Once and Future Story of the Ice Ages*. Berkeley: University of California Press, 2013.

Ruddiman, William F. *Earth's Climate: Past and Future*. 3rd ed. New York: Freeman, 2013.

Woodward, Jamie. *The Ice Age: A Very Short Introduction*. Oxford: Oxford University Press, 2014.

INDEX